SMP interact

C2

Teacher's guide to Book C2

CAMBRIDGE
UNIVERSITY PRESS

PUBLISHED BY THE PRESS SYNDICATE OF THE UNIVERSITY OF CAMBRIDGE
The Pitt Building, Trumpington Street, Cambridge, United Kingdom

CAMBRIDGE UNIVERSITY PRESS
The Edinburgh Building, Cambridge CB2 2RU, UK
40 West 20th Street, New York, NY 10011-4211, USA
477 Williamstown Road, Port Melbourne, VIC 3207, Australia
Ruiz de Alarcón 13, 28014 Madrid, Spain
Dock House, The Waterfront, Cape Town 8001, South Africa

http://www.cambridge.org

First published 2001

Printed in the United Kingdom at the University Press, Cambridge
*Typeface* Minion     *System* QuarkXPress®

A catalogue record for this book is available from the British Library

ISBN 0 521 78539 1     paperback

Typesetting and technical illustrations by The School Mathematics Project
Photographs by Paul Scruton
Cover image © Image Bank/Antonio Rosario
Cover design by Angela Ashton

The maps on pages 67 and 68 are based on Ordnance Survey mapping with the permission of the Controller of
Her Majesty's Stationery Office. © Crown copyright, Licence No. 100001679.

# Contents

The following people contributed to the writing of the SMP Interact key stage 3 materials.

| | | | |
|---|---|---|---|
| Ben Alldred | Ian Edney | John Ling | Susan Shilton |
| Juliette Baldwin | Steve Feller | Carole Martin | Caroline Starkey |
| Simon Baxter | Rose Flower | Peter Moody | Liz Stewart |
| Gill Beeney | John Gardiner | Lorna Mulhern | Pam Turner |
| Roger Beeney | Bob Hartman | Mary Pardoe | Biff Vernon |
| Roger Bentote | Spencer Instone | Peter Ransom | Jo Waddingham |
| Sue Briggs | Liz Jackson | Paul Scruton | Nigel Webb |
| David Cassell | Pamela Leon | Richard Sharpe | Heather West |

Others, too numerous to mention individually, gave valuable advice, particularly by commenting on and trialling draft materials.

| Editorial team: | David Cassell | Project Administrator: | Ann White |
|---|---|---|---|
| | Spencer Instone | Design: | Melanie Bull |
| | John Ling | | Tiffany Passmore |
| | Mary Pardoe | | Martin Smith |
| | Paul Scruton | Project support: | Carol Cole |
| | Susan Shilton | | Pam Keetch |
| | | | Nicky Lake |
| | | | Jane Seaton |
| | | | Cathy Syred |

Special thanks go to Colin Goldsmith.

# Introduction

## What is distinctive about *SMP Interact*?

*SMP Interact* sets out to help teachers use a variety of teaching approaches in order to stimulate pupils and foster their understanding and enjoyment of mathematics.

A central place is given to discussion and other interactive work. Through discussion with the whole class you can find out about pupils' prior understanding when beginning a topic, can check on their progress and can draw ideas together as work comes to an end. Working interactively on some topics in small groups gives pupils, including the less confident, a chance to clarify and justify their own ideas and to build on, or raise objections to, suggestions put forward by others.

Questions that promote effective discussion and activities well suited to group work occur throughout the material.

*SMP Interact* has benefited from extensive and successful trialling in a variety of schools. The practical suggestions contained in the teacher's guides are based on teachers' experiences, often expressed in their own words.

## Who are Books C1 to 3 for?

Books C1 to 3 follow on from Books 1 and N and cover national curriculum levels up to 7. (Level 8 can be covered from Book H1, the first of the higher tier KS4 books.)

## How are the pupils' books intended to be used?

The pupils' books are a resource which can and should be used flexibly. They are not for pupils to work through individually at their own pace. Many of the activities are designed for class or group discussion.

Activities intended to be led by the teacher are shown by a solid strip at the edge of the pupil's page, and a corresponding strip in the margin of the teacher's guide, where they are fully described.

A broken strip at the edge of the page shows an activity or question in the pupil's book that is likely to need teacher intervention and support.

Where the writers have a particular way of working in mind, this is stated (for example, 'for two or more people').

Where there is no indication otherwise, the material is suitable for pupils working on their own.

Starred questions (for example, *C7) are more challenging.

## What use is made of software?

Points at which software (on a computer or a graphic calculator) can be used to provide effective support for the work are indicated by these symbols, referring to a spreadsheet, graph plotter or dynamic geometry package respectively. Other suggestions for software support can be found on the SMP's website: www.smpmaths.org.uk

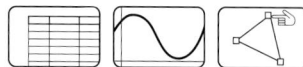

## How is the attainment of pupils assessed?

The interactive class sessions provide much feedback to the teacher about pupils' levels of understanding.

Each unit of work begins with a statement of the key learning objectives and finishes with questions for self-assessment ('What progress have you made?') The latter can be incorporated into a running record of progress.

Revision questions are included in periodic reviews in the pupil's book.

A pack of assessment materials for Books T2, S2 and C2 contains photocopiable masters providing a short assessment for most of the units. Enclosed with the pack is a CD-ROM holding the assessment materials in question bank form so you can compile and edit tests on screen to meet your school's needs. Details of the pack are on the SMP's website.

## What will pupils do for homework?

The practice booklets may be used for homework.

Often a homework can consist of preparatory or follow-up work to an activity in the main pupil's book.

## Answers to questions on resource sheets

For reasons of economy, where pupils have to write their responses on a resource sheet the answers are not always shown in this guide. For convenience in marking you could put the correct responses on a spare copy of each sheet and add it to a file for future use.

# General guidance on teaching approaches

## Getting everyone involved

When you are introducing a new idea or extending an already familiar topic, it is important to get as many pupils as possible actively engaged.

**Posing key questions**
A powerful technique for achieving this is to pose one or two key questions, perhaps in the form of a novel problem to be solved. Ask pupils, working in pairs or small groups, to think about the question and try to come up with an answer.

When everyone has had time to work seriously at the problem (have a further question ready for the faster ones), you can then ask for answers, without at this stage revealing whether they are right or wrong (so that pupils have to keep thinking!). You could ask pupils to comment on each other's answers.

**Open tasks**
Open tasks and questions are often good for getting pupils to think, or thinking more deeply. For example, 'Working in groups of three or four, make up three questions which can be solved using the technique we have just been learning. Try to make your questions as varied as possible.'

## Questioning skills

**Questioning with the whole class**
If your questions to the class are always closed, and you reward the first correct response you get, then you have no way of telling whether other pupils knew the correct answer or whether they had thought about the question at all. It is better to try to get as many pupils as possible to engage with the question, so do not at first say whether an answer is right or wrong. You could ask a pupil how they got their answer, or you could ask a second pupil how they think the first one got their answer.

## Working in groups

**Types of group work**
Group work may be small scale or large scale. In small scale group work, pupils are asked to work in pairs or small groups for a short while, perhaps to come up with a solution to a novel type of problem before their suggestions are compared. In large scale group work, pupils carry out in groups a substantial task such as planning a statistical inquiry or designing a poster to get over the essential idea of the topic they have just been studying.

**Organising the groups**
Group size is important. Groups of more than four or five can lead to some pupils making little or no contribution.

For some activities, you may want pupils to work unassisted. But for many, your own contribution will be vital. Then it is generally more effective if, once you are sure that every group has got started, you work intensively with each group in turn.

**After the group work**    One way to help pupils feel that they have worked effectively is to get them to report their findings to the whole class. This may be done in a number of different ways. One pupil from each group could report back. Or you could question each group in turn. Or each group could make a poster showing their results.

## Managing discussion

Discussion, whether in a whole-class or group setting, has a vital role to play in developing pupils' understanding. It is most fruitful in an atmosphere where pupils know their contributions are valued and are not always judged in terms of immediate correctness. It needs careful management for it to be effective and teachers are often worried that it will get out of hand. Here are a few common worries, and ways of dealing with them.

**What if ...**    '... the group is not used to discussion?'

- Allow time for pupils to work first on the problem individually or in small groups, then they will all have ideas to contribute.

'... everyone tries to talk at once?'

- Set clear rules. For example, pupils raise their hands and you write their name on the board before they can speak.

'...  a few pupils dominate whole-class discussion?'

- Precede any class discussion with small-group discussion and nominate the pupils who will feed back to the class.

'... one pupil reaches the end point of a discussion immediately?'

- Tell them that the rest of the group need to be convinced and ask the pupil to convince the rest of the group.

- Accept the suggestion and ask the rest of the group to comment on it.

# ① Graphs that tell stories

This involves interpretation of the overall features of graphs, particularly the shapes of curves, rather than focusing on the values for particular points.

> **Optional**
> Sheet 191
>
> **Practice booklet** pages 3 to 7

## A **Into the bath** (p 4)

Optional: Sheet 191 for use with question A2

◊ Start by asking the pupils to look at the graph of the water level in Peter's bath. Emphasise that the graph shows the **water level**, **not the volume** of water in the bath. Bring out what each section of the graph implies is happening (see below) but defer discussion of what happens after 30 minutes since question A1 covers this.

*'They did enjoy writing and sharing their stories.'*

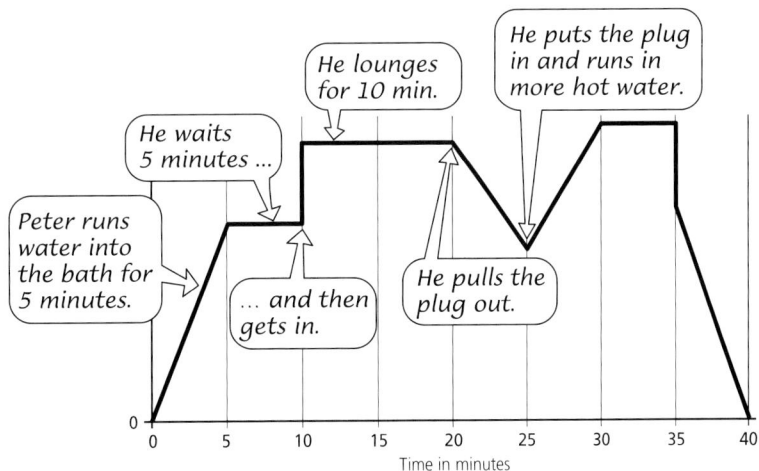

He lounges for 10 min.

He puts the plug in and runs in more hot water.

He waits 5 minutes ...

Peter runs water into the bath for 5 minutes.

... and then gets in.

He pulls the plug out.

Time in minutes

**A2** Some pupils may be helped if they have a copy of sheet 191. They can label events using bubbles as above or can mark key points and refer to them in their stories.

## B **Filling up** (p 6)

◊ In trials, schools found it useful to introduce this section practically with the teacher filling containers from a tap.

Ensure that the flow of water you use is constant, and that the shape of each container allows the varying speed at which the water level changes to be obvious.

## C **Speed** (p 8)

◊ A mistake that pupils frequently make when interpreting a speed–time graph is to think that the shape of the graph reflects the terrain over which the cyclist, for example, is travelling. So they expect that going slowly up a hill is shown by a 'hill' in the graph.

If this problem arises, you could try a point by point approach. Suggest a story such as 'Jay is cycling steadily, but then goes up a steep hill. She then cycles very quickly down the other side.'

Sketch speed–time axes, and plot a first point for Jay's speed.

Point a little further along the time axes, and ask, 'She's cycling steadily. What is her speed compared with before?' Then plot the corresponding point. A little further along the time axis, ask, 'She comes to a steep hill. What's her speed now compared with before?' and again plot the point.

Carry on like this until her whole journey has been discussed, and then join the points.

Another confusion is for pupils to treat a speed graph as if it was a distance graph (for example, a sloping line is interpreted as going at a steady speed).

## A **Into the bath** (p 4)

**A1** (a) Peter is lounging (the water level is constant).

(b) He gets out and pulls out the plug.

(c) The water is running out.

**A2** The pupil's stories corresponding to the three graphs. In the following timings times like 17 minutes are only approximate.

## Nikki's bath

| | |
|---|---|
| 0 to 5 min | runs water |
| at 5 min | gets in |
| 5 to 10 min | lounges |
| 10 to 15 min | runs more water |
| 15 to 17 min | lounges |
| 17 to 20 min | lets out some water |
| 20 to 30 min | lounges |
| at 30 min | gets out and pulls out plug |
| at 35 min | bath is empty |

## Mike's bath

| | |
|---|---|
| 0 to 5 min | runs water |
| at 5 min | gets in and immediately pulls out plug |
| 5 to 7 min | water drains out |
| 7 to 13 min | puts plug back in and runs more water |
| 13 to 20 min | lounges |
| 20 to 23 min | lets out some water |
| 23 to 25 min | puts plug in and runs more water |
| 25 to 30 min | lounges |
| 30 to 35 min | lets out water |
| at 35 min | gets out |
| at $37\frac{1}{2}$ min | bath is empty |

## Chris's bath

| | |
|---|---|
| 0 to 5 min | runs water |
| at 5 min | gets in and almost immediately out! |
| 6 to 10 min | runs more water (probably cold!) |
| at 10 min | gets back in |
| 10 to 20 min | lounges but water is slowly running out |
| at 20 min | puts the plug in properly! |
| 20 to 23 min | lounges |
| 23 to 24 min | runs more water |
| 24 to 30 min | lounges |
| at 30 min | gets out and pulls the plug out |
| at 35 min | bath is empty |

**A3**

*We don't know this time exactly.*

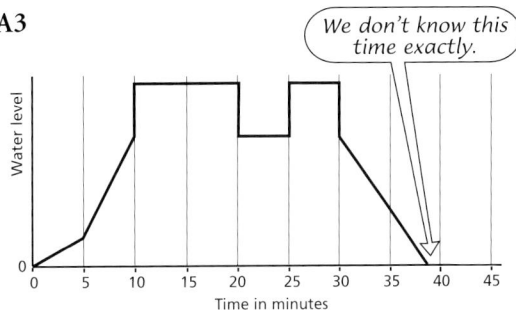

**A4** The pupil's sketch graph and another pupil's story

## B  Filling up (p 6)

**B1** Graph A

**B2** Bottle A goes with graph 2.
Bottle B goes with graph 3.
Bottle C goes with graph 1.

**B3** The pupil's sketch graphs similar to these:

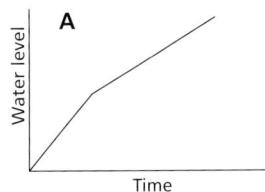

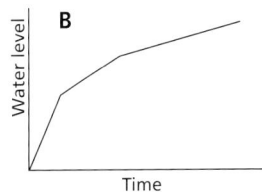

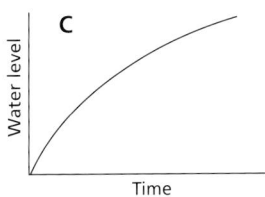

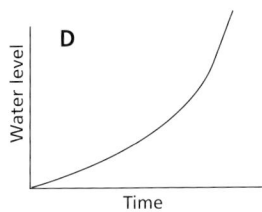

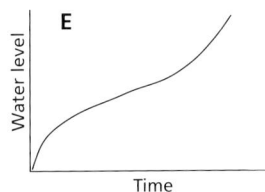

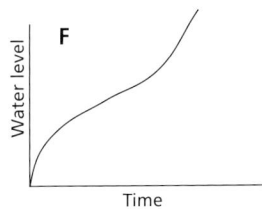

**B4** The pupil's sketches of containers roughly like these:

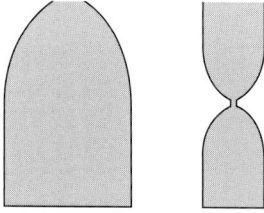

**B5** The pupil's sketch

C **Speed** (p 8)

**C1** (a) Hill View Road

(b) School Road

(c) (i) Clark Road

(ii) 2 minutes

**C2** (a) 50 minutes

(b) Between 30 and 40 minutes

(c) Between 60 and 70 minutes

**C3**

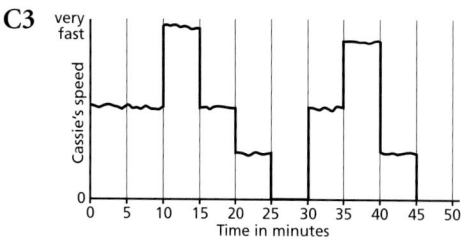

**C4** (a) Graph B

(b) The pupil's stories for A and C

**C5**

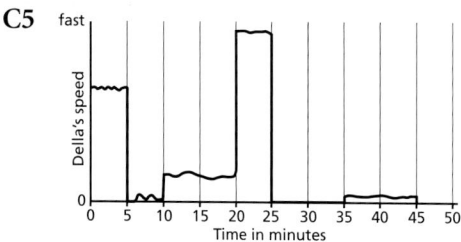

**C6** The pupil's story and graph

**What progress have you made?** (p 11)

1

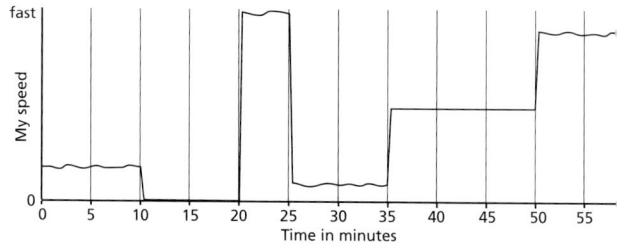

2  0 to 5 min        runs water slowly
   at 5 min          turns off tap
   7 to 12 min       runs water slowly
   12 to 15 min      runs water fast
   15 to 25 min      gets in and lounges
   25 to 30 min      runs water slowly
   30 to 35 min      lounges
   35 to 40 min      lets water out
   at 40 min         gets out
   at 45 min         bath is empty

3  For the first 10 minutes the car
   gradually speeded up.
   Then it went at a steady medium speed
   for 15 minutes.
   It stopped for 5 minutes.
   It speeded up for 5 minutes until it was
   going very fast.
   It continued to go very fast for
   15 minutes.
   For the next 10 minutes it slowed down
   gradually.
   Then it went medium-fast for
   12 minutes, then stopped suddenly.

4

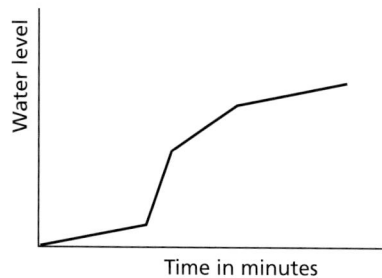

5

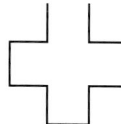

**Practice booklet**

## Section A (p 3)

1 (a) It stays at the 2nd floor for a short time, then goes down to the 1st floor, waits for 20 seconds and then goes down to the basement.

(b) The lift starts to go directly up to the 5th floor.

(c) About 10 seconds

(d) 2 minutes

(e) 1st floor

(f) About 2 minutes 20 seconds

2

| | |
|---|---|
| 0 to 5 min | runs one tap |
| 5 to 10 min | runs other tap as well |
| 10 min | gets in and immediately out |
| 10 to 12 min | plug is out |
| 12 to 20 min | runs in more water |
| at 20 min | gets in again |
| 20 to 30 min | lounges |
| 30 to 32 min | plug is out |
| 32 to 35 min | runs in more water |
| 35 to 40 min | lounges |
| at 40 min | gets out, takes out plug |
| 40 to 47 min | water runs out |

3 (a) 9 a.m.        (b) 8 p.m.

(c) About 5 p.m.   (d) More at 4 p.m.

(e) More coming in

(f) More leaving

*4 (a) At 3 minutes   (b) At 5 minutes

(c) At 8 minutes

(d) (i)  at 6 minutes

(ii) at 14 minutes

(e) 4 people

## Section B (p 5)

1 Graph 2

2 A: graph 2    B: graph 1    C: graph 3

3

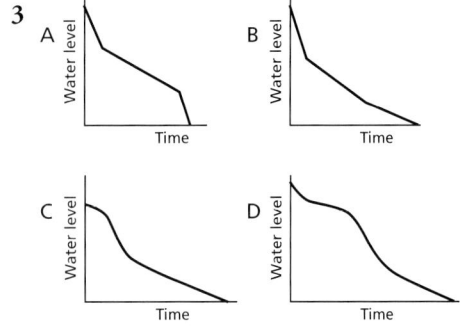

4

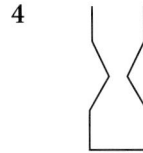

## Section C (p 7)

1 (a) Between 33 and 40 minutes

(b) Between 10 and 20 minutes

(c) Between 20 and 30 minutes

2

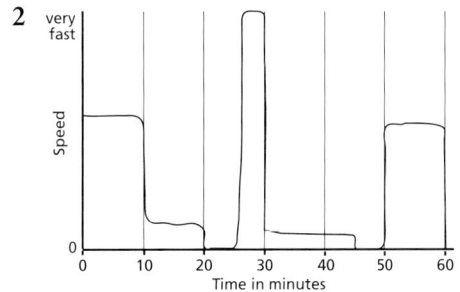

3 (a) Long jump (he speeds up quickly, carries on at the same speed for a short while, and then comes to a dead stop).

(b)

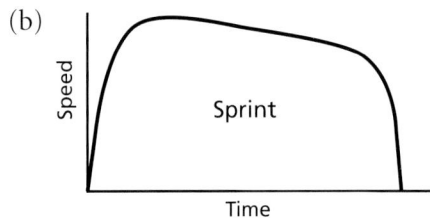

The sprinter fairly quickly gets up to top speed, and then her speed goes down a little bit as she gets tired.

When she reaches the finish line, she fairly quickly slows down to a halt.

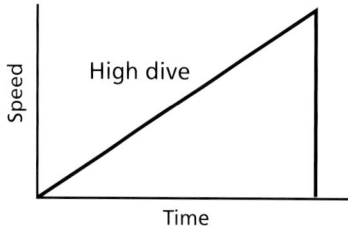

High dive

Speed / Time

The high diver's speed gets steadily faster due to gravity. When he hits the water, his speed goes down to zero very quickly.

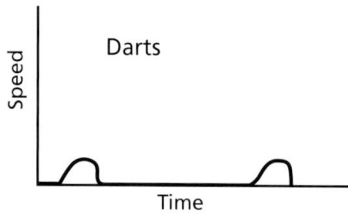

Darts

Speed / Time

The darts player walks to the line. Then she stands there for quite a while until she has thrown her darts. Then she walks to get her darts back.

# ② Scaling

---

**Essential**

Squared paper (preferably centimetre squares)
Pieces of string about 45 cm

**Practice booklet** pages 8 to 12

---

## Ⓐ Spotting enlargements (p 13)

◊ For A1 (the L-shapes) pupils can work in small groups. Encourage them to say why they think particular Ls are or are not enlargements of the one at the top of the page.

## Ⓑ Enlarging a shape (p 14)

Squared paper (preferably centimetre squares)

◊ This provides practice in drawing to a given scale factor. The ideas that all lengths will be multiplied by the scale factor and that angles stay the same should emerge.

**B3** The 'enlarging logos' activity has proved to be an attractive piece of consolidation work in which pupils can use their own creativity.
If pupils have no ideas for a logo they could draw a stylized version of their initials on squared paper.

## Ⓒ Scaling down (p 15)

The term 'scaling down' is used here because many pupils find the idea of a 'half times enlargement' confusing, while a 'half times reduction' falsely implies that that a different mathematical process is being employed when the scale factor is less than 1.

◊ You can ask pupils to identify enlargements or scalings among the shapes (for example, 'B is an enlargement of E with a scale factor 2'). You can leave the search open-ended or say, for example, 'Find me an enlargement with scale factor 3 ... Can you find more than one?' or 'What scale factor enlarges H to A?'

Whenever pupils give an answer, encourage them to say how they got it.

Enlargements by, for example, $1\frac{1}{2}$ and $2\frac{1}{2}$ are relatively easy to spot and understand, but it may be useful to bring in the idea that $1\frac{1}{2}$ can be rewritten as the improper fraction $\frac{3}{2}$, giving a clue that this is to do with finding one-half of a length and then multiplying by 3.

If pupils seem confident enough you can move on to ask, for example, 'What scale factor enlarges A to B?' Some pupils may just 'see' that the answer is $1\frac{2}{3}$. But again it may be a good idea to get them to rewrite this as an improper fraction, relating this to division by 3 followed by multiplication by 5. They should then be in a better position to deal with pairs of shapes where the scale factor is less obvious.

The complete set of scale factors is as follows:

| From \ To | A | B | C | D | E | F | G | H | I | J |
|---|---|---|---|---|---|---|---|---|---|---|
| A | 1 | $\frac{5}{3}$ | $\frac{4}{3}$ | $\frac{7}{6}$ | $\frac{5}{6}$ | $\frac{1}{2}$ | $\frac{1}{3}$ | $\frac{2}{3}$ | 2 | $\frac{7}{3}$ |
| B | $\frac{3}{5}$ | 1 | $\frac{4}{5}$ | $\frac{7}{10}$ | $\frac{1}{2}$ | $\frac{3}{10}$ | $\frac{1}{5}$ | $\frac{2}{5}$ | $\frac{6}{5}$ | $\frac{7}{5}$ |
| C | $\frac{3}{4}$ | $\frac{5}{4}$ | 1 | $\frac{7}{8}$ | $\frac{5}{8}$ | $\frac{3}{8}$ | $\frac{1}{4}$ | $\frac{1}{2}$ | $\frac{3}{2}$ | $\frac{7}{4}$ |
| D | $\frac{6}{7}$ | $\frac{10}{7}$ | $\frac{8}{7}$ | 1 | $\frac{5}{7}$ | $\frac{3}{7}$ | $\frac{2}{7}$ | $\frac{4}{7}$ | $\frac{12}{7}$ | 2 |
| E | $\frac{6}{5}$ | 2 | $\frac{8}{5}$ | $\frac{7}{5}$ | 1 | $\frac{3}{5}$ | $\frac{2}{5}$ | $\frac{4}{5}$ | $\frac{12}{5}$ | $\frac{14}{5}$ |
| F | 2 | $\frac{10}{3}$ | $\frac{8}{3}$ | $\frac{7}{3}$ | $\frac{5}{3}$ | 1 | $\frac{2}{3}$ | $\frac{4}{3}$ | 4 | $\frac{14}{3}$ |
| G | 3 | 5 | 4 | $\frac{7}{2}$ | $\frac{5}{2}$ | $\frac{3}{2}$ | 1 | 2 | 6 | 7 |
| H | $\frac{3}{2}$ | $\frac{5}{2}$ | 2 | $\frac{7}{4}$ | $\frac{5}{4}$ | $\frac{3}{4}$ | $\frac{1}{2}$ | 1 | 3 | $\frac{7}{2}$ |
| I | $\frac{1}{2}$ | $\frac{5}{6}$ | $\frac{2}{3}$ | $\frac{7}{12}$ | $\frac{5}{12}$ | $\frac{1}{4}$ | $\frac{1}{6}$ | $\frac{1}{3}$ | 1 | $\frac{7}{6}$ |
| J | $\frac{3}{7}$ | $\frac{5}{7}$ | $\frac{4}{7}$ | $\frac{1}{2}$ | $\frac{5}{14}$ | $\frac{3}{14}$ | $\frac{1}{7}$ | $\frac{2}{7}$ | $\frac{6}{7}$ | 1 |

## D Scales for maps and drawings (p 16)

Pieces of string about 45 cm long

**D13** This poses something of a challenge and pupils can use various valid ways of reasoning. You could ask pupils to work on it in small groups and then have representatives from a few of the groups explain their reasoning.

## E Ratios (p 20)

◊ Emphasise that ratios involve pure numbers, so that when the quantities to be compared are given in different units, they must first be expressed in the same units.

### A Spotting enlargements (p 13)

**A1** B, F and G are enlargements; the rest are not.

### B Enlarging a shape (p 14)

**B1** (a)  The pupil's enlargement
(b)  They are equal.

**B2** (a)  The pupil's enlargement
(b)  The perimeter is doubled.

**B3** The pupil's logo and enlargement

**B4** The pupil's enlargements

### D Scales for maps and drawings (p 16)

**D1** (a)  About 200 m

**D2** (a)  About 300 m  (b)  About 500 m
(c)  About 125 m  (d)  About 275 m

**D3** (a)  About 2100 m (b)  About 2.1 km

**D4** 50 metres

**D5** 10 metres

**D6** (a)  About 25 by 18 metres
(b)  About 46 metres
(c)  About 33 metres
(d)  About 33 metres

**D7** (a)  3.6 km          (b)  About 110 km

**D8** Empire State Building  380 m
Chrysler Building        320 m
Canada Tower             245 m
Sears Tower              440 m
Petronas Towers          450 m
(These heights include the spires.)

**D9** 1 cm represents 20 m.

**D10** 1 cm represents 40 m.

**D11** (a)  The pupil's scale drawing
(b)  9.0 m

**D12** (a)  9.95 cm          (b)  1.5 cm

**D13** (a)  Map A: 1 cm represents 5 km.
Map C: 1 cm represents 1 km.
Map D: 1 cm represents 2 km.
(b)  Map B: 1 cm represents 10 km.

### E Ratios (p 20)

**E1** (a)  1:50          (b)  1:500
(c)  1:20 000      (d)  1:100 000

**E2** 1:5000

**E3** 1:1000, 1:360 000

**E4** (a)  1:25          (b)  1:50
(c)  1:20          (d)  1:20 000

**E5** (a)  25 m          (b)  About 159 m
(c)  About 113 m

**E6** (a)  2.5 m          (b)  1:250

**E7** 1:400

**E8** (a)  1:6      (b)  1:250      (c)  1:300

### What progress have you made? (p 21)

1

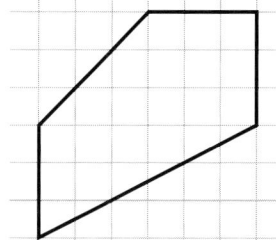

**2** (a)  2          (b)  $\frac{1}{2}$          (c)  $\frac{1}{3}$

 (d)  $\frac{3}{2}$ or $1\frac{1}{2}$   (e)  $\frac{2}{3}$

**3**  17 m,  8 m

**4** (a)  4.8 cm by 3.5 cm

 (b)  120 cm by 87.5 cm

**5** (a)  1 : 200          (b)  1 : 250

 (c)  1 : 5000          (d)  1 : 50 000

## Practice booklet

### Sections A, B and C (p 8)

 **1** The pupil's enlargement

 **2** (a)  The angles of corresponding vertices are the same.

 (b)  It is an enlargement.

 **3** (a)  The angles of corresponding vertices are the same.

 (b)  It is not an enlargement.

 **4** A enlargement of C   scale factor 2

 B enlargement of E   scale factor $\frac{1}{4}$

 B enlargement of G   scale factor $\frac{1}{3}$

 B enlargement of H   scale factor $\frac{1}{2}$

 C enlargement of A   scale factor $\frac{1}{2}$

 E enlargement of B   scale factor 4

 E enlargement of G   scale factor $\frac{4}{3}$ or $1\frac{1}{3}$

 E enlargement of H   scale factor 2

 G enlargement of B   scale factor 3

 G enlargement of E   scale factor $\frac{3}{4}$

 G enlargement of H   scale factor $\frac{3}{2}$ or $1\frac{1}{2}$

 H enlargement of B   scale factor 2

 H enlargement of E   scale factor $\frac{1}{2}$

 H enlargement of G   scale factor $\frac{2}{3}$

 Pupils may not find all the fractional scalings. Shapes D and F are not related by scaling to any other shapes.

 **5** A enlargement of B   scale factor $\frac{3}{2}$ or $1\frac{1}{2}$

 B enlargement of A   scale factor $\frac{2}{3}$

 C enlargement of D   scale factor $\frac{3}{2}$ or $1\frac{1}{2}$

 C enlargement of E   scale factor $\frac{3}{4}$

 D enlargement of C   scale factor $\frac{2}{3}$

 D enlargement of E   scale factor $\frac{1}{2}$

 E enlargement of C   scale factor $\frac{4}{3}$ or $1\frac{1}{3}$

 E enlargement of D   scale factor 2

 **6** (a)  $\frac{1}{2}$          (b)  2

 **7** (a)  $\frac{1}{3}$          (b)  3

### Section D (p 11)

 **1** (a)  7.9 cm          (b)  790 m

 **2** (a)  680 m          (b)  500 m

 (c)  1300 m          (d)  380 m

 **3** The pupil's description of method

 **4** (a)  114 m     (b)  54 m        (c)  79 m

 **5** Scale factor 5

 **6** (a)  14.2 cm          (b)  2.9 cm

### Section E (p 12)

 **1** (a)  1 : 10 000          (b)  1 : 25 000

 (c)  1 : 500 000          (d)  1 : 100

 **2** (a)  1 : 50 000          (b)  1 : 2500

 (c)  1 : 2000

# ③ Graphs and charts

---

**Practice booklet** pages 13 and 14

---

## Ⓐ Off the record (p 22)

◊ Emphasise that in this graph intermediate points have no meaning. The points are joined up to help the eye.

## Ⓑ Drawing graphs and charts (p 23)

This page provides data that you can use for a class lesson on the techniques of drawing line graphs and frequency charts. It is obviously better, if possible, to use data relevant to the class.

**Graphs**

◊ Take the pupils through the steps of drawing a line graph, asking for their suggestions at each stage:
- Find the smallest and largest values for each axis and decide on a scale.
- Draw and label each axis. (If a non-zero start could be misleading, then the 'break' in the scale can be shown by a jagged line.)
- Plot the points; join them up.
- Write the title of the graph.

◊ For the first data set (average daily maximum temperatures) intermediate points have no meaning. For the second, where the process is continuous, interpolation is possible (but is always guesswork).

**Frequency bar charts**

◊ Again, take pupils through the steps involved, inviting their suggestions:
  - Decide on the class intervals to use.
  - Make a table of frequencies.
  - Decide on a scale for each axis.
  - Draw and label the chart and write its title.

◊ For discrete data (as in the case of the pupils' test scores data) class intervals can be marked on the axis with gaps.

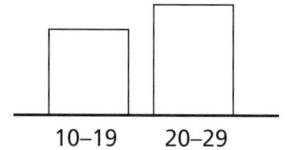

For continuous data (such as the race times), a continuous scale should be used.

The problem about boundary values does not arise in the race times data set. This issue can be left to unit 25, 'Distributions', where pupils are shown how to use inequalities to describe class intervals precisely.

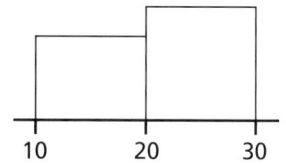

## ℂ **Which graph?** (p 24)

The object of this section is to raise awareness that some graphs and charts may be more appropriate than others, or serve different purposes.

## 𝔻 **Road accidents** (p 26)

Pupils need to choose the appropriate chart(s) to answer each question.

◊ Pupils could be asked to write a newspaper report about accidents in Cornwall using the information in the charts.

### 𝔸 **Off the record** (p 22)

**A1** They fell but less rapidly, becoming almost constant from 1996 to 1998.

**A2** After a slight increase between 1988 and 1989, sales of cassettes fell fairly rapidly between 1989 and 1992. They then fell slightly for three years, then more rapidly.

**A3** (a) Cassettes    (b) CDs

**A4** (a) 1989    (b) 1992

### ℂ **Which graph?** (p 24)

**C1** Sanpam
(looking for highest point on the graph)
(a) Graph A
(b) Cannot tell from B or C

**C2** Sanpam
(comparing the frequency of 22–25)
(a) Graph B
(b) No

**C3** Sanpam
(minimum lower, maximum higher)
(a) Graph A
(b) Cannot tell from B or C

**C4** Sanpam (greater % of hotter ranges)
(a) Graph C
(b) No

**C5** Sanpam temperature went up
Tolero temperature went down
(a) Graph A
(b) Cannot tell from B or C

**C6** Tolero
(two central bars cover a wider range)
(a) Graph C
(b) No

**C7** Tolero's most common range was 14–17.
Sanpam's most common range was
18–21.
(a) Graph B
(b) No

**C8** No, because the lines join one midday
temperature to the next and do not
represent the temperatures in between.

Ⅾ **Road accidents** (p 26)

**D1** About 15%

**D2** About 110

**D3** (a) Going too fast, negligent manoeuvre
(b) 62%

**D4** (a) 10–14    (b) 15–19    (c) 10–14

**\*D5** No. We do not know what proportions of
front and rear seat passengers were
injured. As children most often sit in the
rear seats, one would expect more rear
seat casualties even if both types of seat
were equally dangerous. The 40 front seat
casualties may be a very high proportion
of all the front seat passengers, whereas
the 130 or so rear seat casulaties could be
a very low proportion of all rear seat
passengers.

**D6** The pupil's three questions (with
answers)

### What progress have you made? (p 28)

**1** A possible answer to this question is

| Mark | Frequency |
|------|-----------|
| 10–14 | 3 |
| 15–19 | 7 |
| 20–24 | 7 |
| 25–29 | 6 |
| 30–34 | 6 |
| 35–39 | 6 |
| 40–44 | 4 |
| 45–49 | 1 |

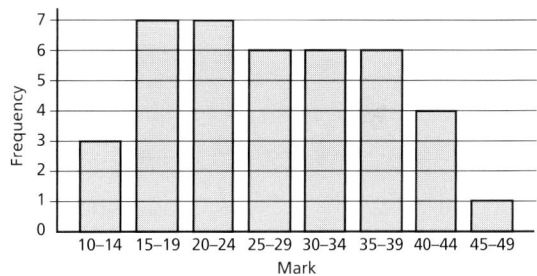

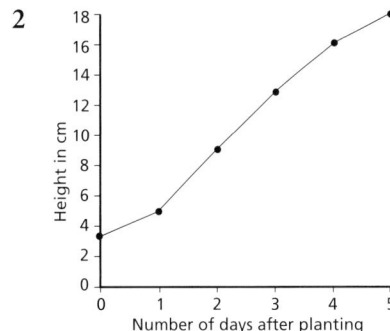

**3** (a) Motorcycle   (b) About 75

(c) 0–14, possibly 20–24, 50+

## Practice booklet

### Section A (p 13)

**1** (a) London, 4°C   (b) Sydney, 22°C

(c) The temperature went down until the 7th and 8th; then went up until the 11th, and then went down again.

(d) On the 4th, 5th, 6th, 7th, 8th, 13th, 14th and 15th

(e) On the 1st, difference 8 degrees

(f) 10°C            (g) On six days

### Section D (p 14)

The pupil's graphs – question 1 frequencies given to help!

**1**

| Weight | Frequency |
|--------|-----------|
| 0–9    | 2         |
| 10–19  | 8         |
| 20–29  | 6         |
| 30–39  | 12        |
| 40–49  | 14        |
| 50–59  | 8         |

# ④ Solving equations

This unit introduces equations of the type $ax - b = cx - d$. Where subtraction is involved, pupils can no longer think of the equations as representing a real-life balance.

Pupils again meet translating number puzzles into equations. This was introduced in 'Think of a number' in *Book C1*. However, the emphasis then was on applying inverses. Inverses cannot be used when the unknown appears in two places in an equation, as it does here.

| | | |
|---|---|---|
| p 29 | **A** Balances review | Solving equations using the balance idea |
| p 30 | **B** Think of a number | Solving number puzzles by translating them into equations |
| p 32 | **C** Both sides | 'Doing the same thing to both sides' |

---

**Optional**

Computer program SOLVE, part of the Slimwam 1 pack available from ATM

**Practice booklet** pages 15 to 17

---

## Ⓐ **Balances review** (p 29)

You will probably wish to take the review on page 29 as a teacher-led activity. Pupils who confidently remember the *Book 1* work will not need to do all the questions in section A.

◊ Talk about 'doing the same thing to both sides' of the balance, rather than 'subtracting the same thing from …' 'Doing the same thing' will be built on in section C.

## Ⓑ **Think of a number** (p 30)

Pupils could discuss in groups how to solve the given puzzles, and then report back to the whole class.

◊ Pupils should realise why they cannot use the inverse approach when solving the second puzzle. The discussion needs to bring out the need to 'do the same thing to both sides' of the equation.

## C **Both sides** (p 32)

(p 32)

This introductory section could be discussed in groups. Then each group could report back on 1 before a class discussion. Discussion on 2 can then follow in a similar way.

> Optional:
> The computer program SOLVE (part of the Slimwam 1 pack available from ATM, tel 01332 346599).
> Graphic calculators such as the TI-92 and computer programs such as Mathematica can solve equations. However, their syntax may not be what you wish to encourage; for example $(4x - 3 = 3x + 5) - 3x$ for 'take $3x$ off both sides'.

◊ In 1, if we wish to isolate the $5n$ on the left-hand side of $5n - 16 = n + 4$, then we must add 16 to both sides.

Pupils' experiences in 'Think of a number' should help them see that in order to get back from $5n - 16$ to $n$, they have first to 'undo' the $-16$.

In solving equations of this type, pupils often think that they can 'take off 16' from $5n - 16$, leaving just $5n$. Emphasising 'How do you undo subtract 16? You add 16' may help, but pupils themselves need to confront this thorny problem.

◊ If the point does not naturally arise in discussion, bring out that there is no one correct thing to do first. In solving $5n - 16 = n + 4$ you could first add 16 to both sides or subtract 4 from both sides with equal validity.

You may prefer pupils to tackle the $n$s first, so that in later work they do not get left with any $^-n$s.

◊ There are innumerable pitfalls when solving equations. The second discussion point may be used to bring out some common mistakes, such as taking $a$ off $a - 6$, and leaving just 6.

In equations with two subtractions, such as $9z - 7 = 4z - 5$, the method of C2 will involve dealing with negative numbers.

**C3** This question attempts to draw pupils' attention to some common errors.

**C10/11** This activity could be played in pairs as a game. Each player takes two cards. The first to solve their equation gains a point.

You may need more cards to make the game interesting. Designing them so that the resulting pairs of equations all have reasonable answers could provide a challenge for older pupils.

## A Balances review (p 29)

**A1** (a) 22 (b) 4 (c) 2 (d) 225

**A2** (a) $t = 4$ (b) $n = 7\frac{1}{2}$
(c) $d = 10$ (d) $x = 5$
(e) $q = 6\frac{1}{2}$ (f) $n = 10$
(g) $m = 0.75$ (h) $n = 3.6$
(i) $e = 1\frac{1}{2}$ (j) $a = 2$
(k) $h = 2$ (l) $x = 42$

**A3** (a) $d = 3$ (b) $b = 6$
(c) $x = 6$ (d) $k = 4$
(e) $g = 3\frac{1}{2}$ (f) $h = 5$
(g) $k = 3$ (h) $m = 3$
(i) $y = 1$ (j) $h = 4$

**A4** The pupil's equation with solution $t = 5$

## B Think of a number (p 30)

Except for B3, pupils will use their own letters in these questions.
In questions B1 and B2 we have used $n$.

**B1** Equation $6n + 20 = 10n$; solution $n = 5$

**B2** (a) $9n + 18 = 11n$, $n = 9$
(b) $4n + 100 = 8n$, $n = 25$

**B3** $3(d + 10) = 42$
$3d + 30 = 42$
$3d = 12$
$d = 4$

**B4** Jay thought of 15.

**B5** (a) Hamish thought of 10.
(b) Anne thought of 5.

**B6** They both started with 4.

**B7** 9

**B8** 4

**B9** 3

**\*B10** Kirsty is 16.

## C Both sides (p 32)

**C1**
$s + 1 = 3s - 17$
$s + 18 = 3s$
$18 = 2s$
$9 = s$

**C2**

**C3** (a) Second line should be $4d = 2d + 8$, so $d = 4$.
She should have added 2 to both sides.

(b) Second line should be $^-2 = 2s - 8$, so $s = 3$.
If you subtract $3s$ from $3s - 2$ you are left with $^-2$.

(c) Second line is incorrect; subtracting 4 from both sides would give $5h - 14 = 4h - 4$, which is no use anyway.
Correct solution is $h = 10$.

**C4** (a) $n = 6$ (b) $b = 4$
(c) $f = 8$ (d) $w = 11$
(e) $y = 3$ (f) $s = 3\frac{1}{2}$
(g) $u = 20$ (h) $w = 17$
(i) $y = 8$ (j) $g = 1\frac{1}{2}$
(k) $p = 19$ (l) $g = 1$
(m) $f = 100$ (n) $a = 28$

**C5** (a) $a = 14$ (b) $x = 16$
(c) $y = 11$ (d) $b = 9$
(e) $j = 12$ (f) $t = 11\frac{1}{2}$
(g) $x = 35$ (h) $b = 7$
(i) $j = 6$ (j) $b = 4\frac{1}{2}$
(k) $t = 8$ (l) $k = 4.8$
(m) $s = 6$ (n) $z = ^-2$

**C6** 12

**C7** Jenny and Bob are 17, Uncle Fred is 87.

**C8** Andy is 12, Barrie is 32, Chris is 36.

**C9** It is February (and today is the 23rd).

**C10** (a) $w = 10$     value of cards = 39

      (b) $w = 10$     value of cards = 41

**C11** Picking $4w - 1 = 6w - 19$, $w = 9$
Pair left over is $3w + 11 = 2w + 19$, $w = 8$

Picking $4w - 1 = 3w + 11$, $w = 12$
Pair left over is $6w - 19 = 2w + 19$, $w = 9\frac{1}{2}$

**C12** As C11

**\*C13** (a) $x = 6$

      (b) $x + 27$, $2x + 21$, $7x - 9$

### What progress have you made? (p 35)

**1** (a) $y = 7$    (b) $f = 5$    (c) $t = 3$

**2** (a) $h = 7$    (b) $r = 3$    (c) $s = 4$

**3** (a) $g = 5$    (b) $x = 7$    (c) $y = 4$

**4** (a) John thinks of 12.

     (b) Viv thinks of 32.

     (c) They both think of $2\frac{1}{2}$.

## Practice booklet

## Section A (p 15)

**1** (a) $p = 9$

     (b) $n = 9$

     (c) $c = 2\frac{1}{2}$

     (d) $x = 3$

     (e) $y = 2$

     (f) $a = 2$

     (g) $m = 0.5$

     (h) $w = 2$

     (i) $h = 5$

     (j) $d = 40$

**2** The pupil's equation

## Section B (p 15)

**1** (a) $7x + 21 = 10x$      $x = 7$

     (b) $5(x + 4) = 50$      $x = 6$

     (c) $3(x + 2) + 10 = 5x$   $x = 8$

     (d) $\frac{x}{2} + 10 = 3x$      $x = 4$

**2** 6

**3** 3

## Section C (p 16)

**1** (a) $n = 3$    (b) $b = 4$    (c) $y = 4$

     (d) $f = 36$    (e) $x = 12$    (f) $p = 1\frac{1}{2}$

     (g) $m = 22$    (h) $w = 10$    (i) $x = 12$

     (j) $y = 14$    (k) $a = 5$    (l) $b = 4$

**2** 7

**3** 10 cm, 7 cm
The answers can be found by solving
$6(x - 3) = 4x + 2$

**4** $x = 1.5$
The answer can be found by solving
$\frac{5x}{2} = \frac{3(x + 1)}{2}$

**5** $w = 6$
The answer can be found by solving
$5w = 3(w + 4)$

**6** $d = 5$
The answer can be found by solving
$4d + 3(d - 2) = 29$

# ⑤ Solids

The aim of the first part of this unit is to help pupils visualise sets of cross-sections of familiar solid objects. Later in the course the idea of a uniform cross-section plays a part in calculating volumes of prisms.

The second part deals with reflection symmetry in three dimensions.

| Essential | Optional |
|---|---|
| A selection of objects similar to those shown in the pupil's book | Water in a suitable container (a three-litre plastic fizzy drinks bottle, cut off below its 'shoulders', works well) |
| | Potatoes, carrots or similar and a sharp knife |
| | Plasticine |
| | Cubes and tightly fitting rubber bands |
| **Practice booklet** pages 18 and 19 | |

## A Cross-sections (p 36)

◊ Use the photos and drawings of the screwdriver to discuss what a cross-section is. Emphasise the idea of cutting through the object at the water level. Many teachers have found cutting through vegetables an effective approach at this stage. Shapes causing difficulty can be reproduced as Plasticine models that are then sliced through.

Some pupils may need guidance about how accurate their sketches need to be. Explain that exact drawings are not required, but try to get over the idea that each sequence of sketches should show the relative sizes of the cross-sections of a particular object.

**A4** After doing this pupils can draw a set of cross-sections of an object of their own choice.

**A5** The set of cross-sections labelled A can lead to a more general discussion of solids which, when lowered into the water a certain way round, produce a cross-section that doesn't change. Can pupils think of some?

**A8** Pupils find the chair difficult, partly because they are not expecting anything quite so big. You may need to give a hint or gesture that 'it's about this big'.

**\*A9** Pupils may find it helpful to have a cube with a rubber band which they can fit round the cube to show the edge of a cut through the cube.

*'To start the lesson I asked for a volunteer… I then brought out my saw and pretended to "cut in half" at knee level and asked [the] class to draw the cross-section… The next cut was at elbow level… the next cut across the shoulders. The final cut was through the head. There was no shortage of volunteers.'*

## B  Planes of symmetry (p 42)

**T**

◊ There are various ways of giving a practical demonstration. If you have a large enough mirror you can place a solid against it and observe that the 'object' consisting of the solid and its reflection has a plane of symmetry – the mirror. Alternatively, you can use objects that can be cut (for example, Plasticine models).

◊ Emphasise that the plane of symmetry is the whole (infinite) plane and not just the cross-section of the solid.

## A  Cross-sections (p 36)

**A1**  C, A, B then D

**A2**  B, C, D then A

**A3**  Trowel    Hoe    Mallet

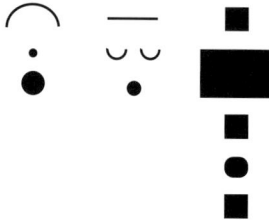

**A4**  C, A, D, B then E

**A5**  1 B, 2 C, 3 A

**A6**  (a)        (b)

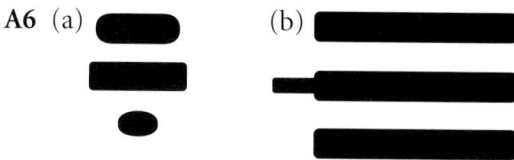

**A7**  1 A, 2 B, 3 A  4 C, 5 A, 6 B

**A8**  (a)  (i)  Fork    (ii)  Chair

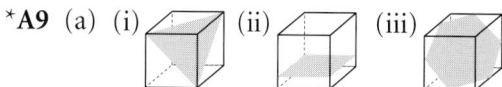

*****A9**  (a)  (i)   (ii)   (iii) 

(b)  There are many (infinitely many) different cross-sections of a cube. Here are some of the more obvious infinite sets of them.

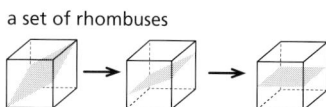

a set of rhombuses

a set of rectangles

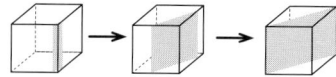

another set of rectangles

a set of isosceles triangles

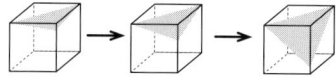

a set of trapezia

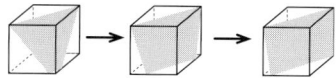

## B  Planes of symmetry (p 42)

**B1**  4

**B2**  4

**B3**  3

**B4**  One possibility is shown for each.

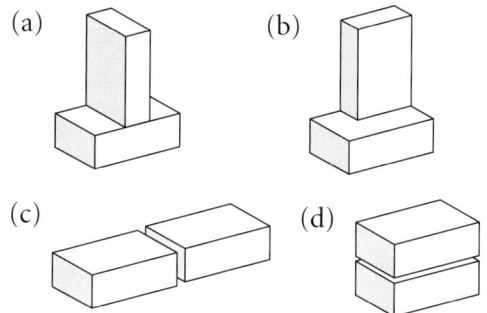

(a)         (b) 

(c)         (d) 

**B5**  9

## What progress have you made? (p 43)

1  E, C, A, B, F, D  or  B, E, C, A, F, D

2  (a)

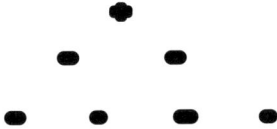

(b)

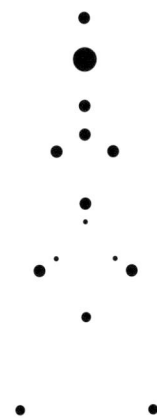

3  5

## Practice booklet

### Section A (p 18)

1  Set 1: slotted spoon  D, C, A, B, E
   Set 2: spatula  A, D, E, C, B
   (A, D and E could be in any order.)
   Set 3: fish slice  C, D, A, E, B

2

3  (a)

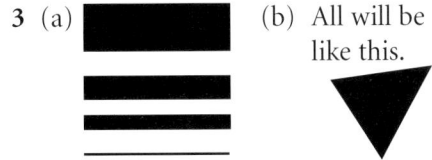

(b)  All will be like this.

4  Some possible drawings:

Pencil          Safety pin

Padlock

5  A sphere

### Section B (p 19)

1  (a)  1          (b)  4

# ⑥ Units

This covers the use of metric units of area, volume and capacity and the relationships between them together with metric equivalents of common imperial units.

---

**Practice booklet** pages 20 and 21

---

## Ⓐ Area (p 44)

◊ This section tackles the common misconception that there are 100 cm² in a square metre.

## Ⓑ Capacity and volume (p 45)

◊ A practical demonstration of the fact that 1 litre = 1000 cm³ can be made using a hollow cube 10 cm by 10 cm by 10 cm and filling it with rice from a 1 litre jug.

The cube is easily made by folding a 30 cm square of card or paper.

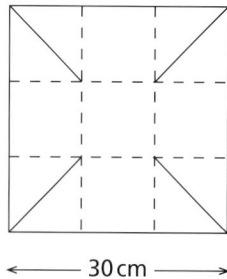

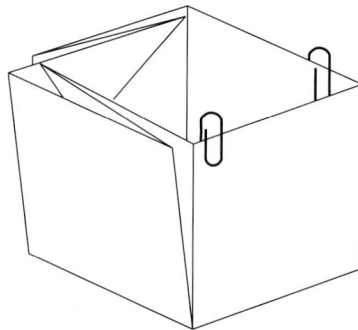

◊ You could ask pupils to guess how deep the juice is in a 1 litre carton, and how they might find this out by measuring the edges of the carton.

◊ The 'design a carton' activity is easily carried out on a spreadsheet. The minimum surface area for a cuboid is given by a cube (edges 7.94 cm).

## Ⓒ Imperial units (p 47)

◊ The units included here are miles, feet, pounds, pints and gallons.

## Ⓓ Mixed questions (p 48)

## Ⓐ Area (p 44)

**A1** $1\,m^2 = 10\,000\,cm^2$ ($1\,m^2$ is the area of a square 100 cm by 100 cm.)

**A2** (a) $8400\,cm^2$     (b) $0.84\,m^2$

**A3** (a) $0.8\,m^2$; $8000\,cm^2$

    (b) $0.7\,m^2$; $7000\,cm^2$

    (c) $2\,m^2$; $20\,000\,cm^2$

**A4** 20 litres

**A5** (a) X: $6000\,cm^2$ or $0.6\,m^2$
      Y: $4000\,cm^2$ or $0.4\,m^2$

    (b) Y, because although X is $1\frac{1}{2}$ times the area of Y, it is more than $1\frac{1}{2}$ times the price (or equivalent reason)

## Ⓑ Capacity and volume (p 45)

**B1** (a) 2 litres     (b) 9 litres
    (c) 20 litres     (d) 672 litres

**B2** (a) 2 litres     (b) 3.5 litres
    (c) 0.85 litres

**B3** (a) 0.25 litres     (b) 0.001 litres
    (c) 0.05 litres     (d) 0.75 litres

**B4** $1\,m^3 = 1\,000\,000\,cm^3$ ($1\,m^3$ is the volume of a cube 100 cm by 100 cm by 100 cm.)

**B5**

| | Capacity | AFF | IAS | TAS |
|---|---|---|---|---|
| (a) | 201.6 litres | 4 tsp | 8 tsp | 100 ml |
| (b) | 2560 litres | 51 tsp | 102 tsp | 1280 ml |
| (c) | 1260 litres | 25 tsp | 50 tsp | 630 ml |

**B6** (a) 1000 litres
    (b) 83 cm

**B7** (a) 120 000 litres
    (b) 7 days (6 days and 23 hours)
    (c) 3 kg or 3000 g

**B8** 8.94 cm (to 2 d.p.)

## Ⓒ Imperial units (p 47)

**C1** Calais 10     Boulogne 15
    Firenze 40     Siena 20
    Limerick 33     Ennis 13

**C2** 30 m.p.h. is roughly 48 km/h.
    50 m.p.h. is roughly 80 km/h.
    70 m.p.h. is roughly 112 km/h.

**C3** (a) 36 litres     (b) 90 litres
    (c) 225 litres     (d) 900 litres

**C4** (a) 12.6 litres     (b) 27 litres
    (c) 72 litres

**C5** 19.8 metres (or 20 metres)

**C6** (a) To change from km to miles, divide by 8 and multiply by 5.

    (b) (i) 30 miles     (ii) 125 miles
        (iii) 37.5 miles     (iv) 90 miles

**C7** (a) 3 metres (b) $9\,m^2$     (c) 11 sq ft

## Ⓓ Mixed questions (p 48)

**D1** 12

**D2** 3.31 kg

**D3** 6000

**D4** 25 days

**D5** 6

**D6** 81 kg

**D7** About 4.5 to 5 litres

**D8** About 82 kg

**D9** Length 60 cm, waist 52.5 cm

### What progress have you made? (p 48)

**1** (a) 2.3 litres     (b) 0.25 litres
    (c) 11.25 litres

**2** Volume = 289 litres
    Takes 24 minutes to fill

**3** (a) 160 km     (b) 6.4 kg

**Practice booklet**

**Sections A and B** (p 20)

1 (a) $1.33\,\text{m}^2$; $13\,300\,\text{cm}^2$
  (b) $0.78\,\text{m}^2$; $7800\,\text{cm}^2$
  (c) $1.74\,\text{m}^2$; $17\,400\,\text{cm}^2$

2 75 tiles

3 (a) 0.1 litres     (b) 0.005 litres
  (c) 1.043 litres   (d) 0.65 litres
  (e) 0.25 litres    (f) 5000 litres

4 (a) 2 litres      (b) 0.1 litres
  (c) 6.48 litres    (d) 216 litres

5 (a) $375\,\text{m}^3$       (b) 375 000 litres
  (c) 8 hours

**Sections C and D** (p 21)

1 (a) 64 km     (b) 168 km    (c) 408 km

2 200 lengths

3 About 9 gallons

4 About 44 pounds

5 25 loads

6 1.4 m = 140 cm
  5 feet is about 150 cm so Rani is taller by
  about 10 cm.

7 16 inch collar

# 7 Simplifying expressions

| Optional |
|---|
| Sheet 192 |

**Practice booklet** pages 22 to 26

## A Simplifying (p 49)

One way to structure the introduction is described below.

◊ Begin by showing pupils a set of statements such as

$3x + 4 = 7x$ $\qquad\qquad$ $3x + 4x = 7x$ $\qquad\qquad$ $3x + 4 = 3x$

and pupils classify them as 'always true', 'sometimes true' or 'never true'.

Establish that $3x + 4 = 7x$ when $x = 1$ but $3x + 4x = 7x$ is true for any value of $x$. Explain that $3x + 4x$ and $7x$ are called equivalent expressions as they have the same value whatever the value of $x$.

◊ Ask pupils to simplify where possible a set of expressions such as:

$x + 4 + 1$ $\qquad\qquad$ $6 - 5x$ $\qquad\qquad$ $7x + 5x + 11$

$9x - 7x$ $\qquad\qquad$ $10x + 7$

If pupils make mistakes, substitute suitable numbers to identify them.
For example $10x + 7 \neq 17x$ because when $x = 2$, $10x + 7 = 27$ but $17x = 34$.

◊ Now ask pupils to try to find any pair of equivalent expressions in set 1 on page 49 of the pupil's book (B and D are equivalent).
It may be helpful to emphasise the following points:

• Trying a few values does not prove equivalence although it suggests it. It is certainly one way to show that two expressions are **not** equivalent (e.g. $5 - x$ and $2x + 5 - x$).

• In calculations using addition and subtraction, order does not matter.
For example: $7 + 5 - 1 - 3 = 7 - 3 + 5 - 1 = 7 - 1 - 3 + 5$
Similarly: $\quad 6 + 2x - 1 - 3x = 6 - 1 + 2x - 3x = 5 - x$

◊ You could show, for example, $5 + 2x - 3x = 5 + 2x + {}^-3x = 5 + {}^-x = 5 - x$. Familiarity with this sort of manipulation can prove very useful.

◊ Now pupils can try to find pairs of equivalent expressions from set 2.

◊ Some misconceptions can be addressed, e.g. $6 - 5y + 2y \neq 6 - 7y$

◊ Ask pupils to find any pairs of equivalent expressions in set 3 and discuss their results.

Ensure pupils see that, for example, $x + 6y$ cannot be simplified further.

## B Magic squares (p 50)

◊ Some questions involve solving simple equations. You may wish to revise techniques for solving equations before pupils start section B.

\*B9 For part (b), since the total is $12a$, any square where $a = 2$ will give a total of 24. Hence, there are an infinite number of solutions.

## C Walls (p 52)

> Optional: Sheet 192 containing blank walls

The context for this section is 'walls' where the number on each 'brick' is found by adding the numbers on the two bricks below.

◊ Ask pupils to look at the first wall or demonstrate a complete wall on the board/OHP transparency. Can pupils see how it has been constructed?

In pairs or groups, pupils try to find the values of $a$ and $b$ for the walls at the top of page 52. To find $a$ they can simply complete the wall but finding $b$ is harder. Some pupils will probably use trial and improvement. Discuss their methods. By the end of the discussion, pupils should understand how to use algebra to form an equation to find the value of $b$ as shown below.

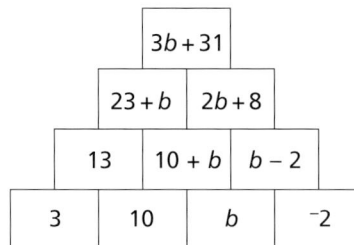

Since the number in the top brick is 100 we know that $3b + 31 = 100$ which can be solved to give $b = 23$.

Check the result by constructing the whole wall to give:

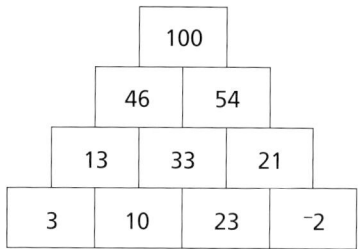

|  |  | 100 |  |  |
|---|---|---|---|---|
|  | 46 |  | 54 |  |
|  | 13 | 33 | 21 |  |
| 3 | 10 | 23 | -2 |  |

◊ You may wish to show that the numbers in the wall could have been found by labelling a different brick.
For example:

|  | 100 |  |
|---|---|---|
|  |  |  |
|  | $x$ |  |
| 3 | 10 | -2 |

|  | $3x + 1$ |  |  |
|---|---|---|---|
|  | $13 + x$ | $2x - 12$ |  |
|  | 13 | $x$ | $x - 12$ |
| 3 | 10 | $x - 10$ | -2 |

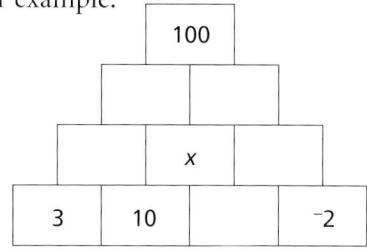

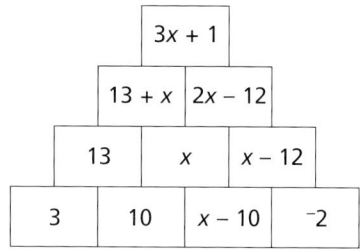

This leads to $3x + 1 = 100$ and $x = 33$ giving, of course, the same final completed wall.

◊ Emphasise to pupils that they should use algebraic techniques to solve the problems in section C and not trial and improvement.

**C1** In part (c), pupils do not need to complete the whole wall to find $z$.

'Worked very well. Some started on trial and improvement on C2 but quickly saw how tedious that was and ... decided that it was very much better using algebra. Very good for showing a reason for using algebra.'

Ⓐ **Simplifying** (p 49)

**A1** (a) $4p$    (b) $5q$    (c) $r + 13$
     (d) $7s + 3$    (e) $6t - 3$    (f) $2u$
     (g) $3v - 1$    (h) $11 + w$    (i) $3x$

**A2** A and G, B and E, C and H, D and F

**A3** (a) $5 - 2k$    (b) $5j - 3$    (c) $1 - h$
     (d) $10 + 2g$    (e) $7 - f$    (f) $6 - 3e$
     (g) $6d - 4$    (h) $2 - 3c$    (i) $7 - 3b$

**A4** (a) $4z + 8y$    (b) $3x + 2w$
     (c) $7 + 4u + 2v$    (d) $3t - s$
     (e) $2 + 5q - 5r$    (f) $5n + 3p$
     (g) $6 + l - 3m$    (h) $10 - 8k - 3j$
     (i) $4h - 4g$    (j) $3e - 4f - 3$
     (k) $3c - 3 - d$    (l) $8a - 4b$

Ⓑ **Magic squares** (p 50)

**B1** (a)

| 31 | 10 | 22 |
|---|---|---|
| 12 | 21 | 30 |
| 20 | 32 | 11 |

(b)

| 8 | 15 | 5 | 14 |
|---|---|---|---|
| 21 | 9 | 3 | 9 |
| 1 | 13 | 17 | 11 |
| 12 | 5 | 17 | 8 |

**B2** Change the 8 to a 5 to turn the grid into a magic square.

**B3** (a)

| 9 | 6 | 6 |
|---|---|---|
| 4 | 7 | 10 |
| 8 | 8 | 5 |

(b)

| 11 | 5 | 7 |
|----|---|---|
| 7 | 8 | 15 |
| 7 | 9 | 7 |

(c) The square in (a) is a magic square.

**B4** (a) $w + 2 + 3w + 8 + 5 - w = 15 + 3w$

(b) Each set of three expressions adds to give $15 + 3w$.

(c) Each total is the same so any value of $w$ will make a magic square.

(d) The magic total is 18.

**B5** (a) $5 + w = 8$ gives $w = 3$ making the magic square below.

| 5 | 17 | 2 |
|---|----|---|
| 5 | 8 | 11 |
| 14 | $^-1$ | 11 |

(b) $15 + 3w = 30$ gives $w = 5$ making the magic square below.

| 7 | 23 | 0 |
|---|----|---|
| 3 | 10 | 17 |
| 20 | $^-3$ | 13 |

**B6** (a)

| $2x + 1$ | $2 + x$ | $6x - 3$ |
|----------|---------|----------|
| **7x – 4** | **3x** | **4 – x** |
| **3** | **5x – 2** | **4x – 1** |

(b) $9x = 18$ gives $x = 2$ making the magic square below.

| 5 | 4 | 9 |
|---|---|---|
| 10 | 6 | 2 |
| 3 | 8 | 7 |

**B7** (a)

| **8 – 3y** | $3 - 4y$ | **4 + y** |
|-----------|----------|-----------|
| **1 + 2y** | $5 - 2y$ | **9 – 6y** |
| **6 – 5y** | $7$ | $2 - y$ |

(b) The total is 9.

**B8** (a)

| **2y + 4z** | $y - z$ | $6y$ |
|------------|---------|------|
| **7y – 3z** | $3y + z$ | $5z - y$ |
| **2z** | **5y + 3z** | $4y - 2z$ |

(b)

| 18 | 3 | 30 |
|----|---|----|
| 29 | 17 | 5 |
| 4 | 31 | 16 |

| $5a+11b-1$ | $16b+4a-7$ | $7-20b$ | $1+3a-7b$ |
|---|---|---|---|
| $7+4a$ | $1-16b$ | $5a+16b-7$ | $3a-1$ |
| $2a-2b-6$ | $3a+6$ | $4a$ | $3a+2b$ |
| $a-9b$ | $5a$ | $4b+3a$ | $3a+5b$ |

(b)  The pupil's squares with $a=2$
Examples are:

| 20 | 17 | -13 | 0 |
|---|---|---|---|
| 15 | -15 | 19 | 5 |
| -4 | 12 | 8 | 8 |
| -7 | 10 | 10 | 11 |

$b=1$

| 9 | 1 | 7 | 7 |
|---|---|---|---|
| 15 | 1 | 3 | 5 |
| -2 | 12 | 8 | 6 |
| 2 | 10 | 6 | 6 |

$b=0$

*B10  (a)  To show that grid B is not a magic square, you need to show that two of its totals are not the same.

The total for the fourth row and for the third column is $33c-2d$.
All the other totals are $34c-6d$.

(b)  Change $12c-d$ (bottom row) to $13c-5d$.

(c)  10

ℂ **Walls** (p 52)

**C1**  (a)  $x=24$     (b)  $y=0.5$     (c)  $z=12$

**C2**  (a)

```
            241
        155     86
      76    79     7
    9    67    12    -5
```

(b)

```
            23
        13      10
      9     4      6
    2     7    -3     9
```

(c)

```
            24
        14.5    9.5
      7.5    7     2.5
    5    2.5   4.5   -2
  4    1    1.5    3    -5
```

(d)

```
              5
          0       5
        3    -3      8
      9    -6     3     5
   14   -5    -1     4     1
```

**C3**  Each number on the bottom row is 2.5 or $2\frac{1}{2}$.

*C4 (a)

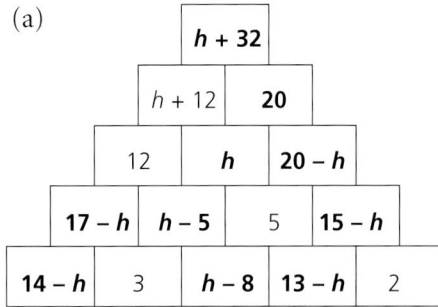

(b) $h = 68$

*C5 $n = 4$

*C6 $y = 1.5$

### What progress have you made? (p 54)

1 (a) $4p$      (b) $5 + 6q$
  (c) $5 - 2r$      (d) $4s - 1$
  (e) $2w - 2x$      (f) $4y - 1 - 2z$

2 (a) (i)

| | | |
|---|---|---|
| $5x + 2$ | 3 | $7x - 2$ |
| $6x - 3$ | **4x + 1** | **2x + 5** |
| **x + 4** | **8x - 1** | **3x** |

(ii)

| | | |
|---|---|---|
| $7a - 2b$ | $2a - 3b$ | **3a + 2b** |
| **3b** | $4a - b$ | **8a - 5b** |
| **5a - 4b** | $6a + b$ | **a** |

(b) The pupil's magic square with 9 in the centre square and their working

3

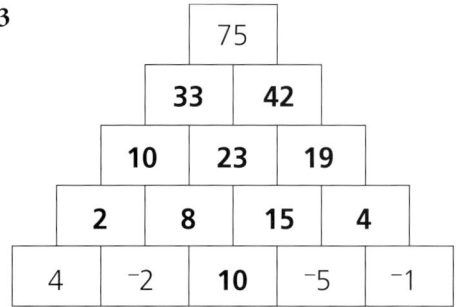

### Practice booklet

### Section A (p 22)

1 D $(3p + 11)$

2 J and O, K and L, M and Q, N and P

3 (a) $6k$      (b) $12 + n$
  (c) $7 - n$      (d) $1 + 5p$
  (e) $5 - 3n$      (f) $3r - 8$
  (g) $4$      (h) $3t$
  (i) $7n - 7$      (j) $v - 2$

4 (a) $7u + 12v$      (b) $11y - x$
  (c) $4w - 8x - 4$      (d) $6a - 3b - 3$
  (e) $4c - 2d + 5$      (f) $2e - 2f - 6$
  (g) $8g - 5h + 3$      (h) $5 - k - 2m$
  (i) $4n - 3p - 2$      (j) $11 - 2p$

### Section B (p 23)

1 (a) (i) The total in each direction is $3a + 21$.

    (ii) Each total simplifies to the same expression so **any** value of $a$ will give a magic square.

(b) (i)

| 4 | 14 | 6 |
|---|---|---|
| 10 | 8 | 6 |
| 10 | 2 | 12 |

(ii)

| -2 | 23 | 12 |
|----|----|----|
| 25 | 11 | -3 |
| 10 | -1 | 24 |

(c) $a + 7 = 12$ gives $a = 5$.

(d) $3a + 21 = 27$ gives $a = 2$ making the magic square below.

| 2  | 17 | 8  |
|----|----|----|
| 15 | 9  | 3  |
| 10 | 1  | 16 |

2 (a) Grid B totals
Row 1: $18b + 9$
Row 2: $18b + 9$
Row 3: $18b + 10$
Column 1: $18b + 9$
Column 2: $18b + 10$
Column 3: $18b + 9$
Both diagonals: $18b + 9$

Grid C totals
$21a - 3b$ in every direction.

(b) Grid C is a magic square.

3 (a) The totals in each direction are $6d + 21$.

(b) (i)

| 4  | 13 | 16 |
|----|----|----|
| 23 | 11 | -1 |
| 6  | 9  | 18 |

(ii)

| 2  | 21 | 22 |
|----|----|----|
| 35 | 15 | -5 |
| 8  | 9  | 28 |

4 Grid E

| $3e + 2$  | $8e + 7$  | $7e$      |
|-----------|-----------|-----------|
| $10e + 1$ | $6e + 3$  | $2e + 5$  |
| $5e + 6$  | $4e - 1$  | $9e + 4$  |

Grid F

| $3f + g$  | $10f + 6g$ | $5f - g$  |
|-----------|------------|-----------|
| $8f$      | $6f + 2g$  | $4f + 4g$ |
| $7f + 5g$ | $2f - 2g$  | $9f + 3g$ |

Grid G

| $4g - 1$ | $4 - g$  | $6g - 3$ |
|----------|----------|----------|
| $5g - 2$ | $3g$     | $g + 2$  |
| $3$      | $7g - 4$ | $2g + 1$ |

Grid H

| $6h - 4j$ | $h + 3j$  | $8h - 2j$ |
|-----------|-----------|-----------|
| $7h + j$  | $5h - j$  | $3h - 3j$ |
| $2h$      | $9h - 5j$ | $4h + 2j$ |

**Section C** (p 25)

1 (a)

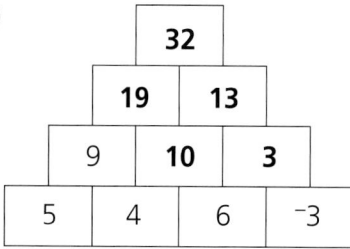

(b)

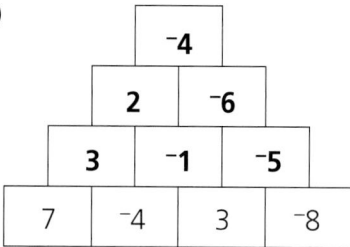

2 (a) $p = 7$      (b) $q = 10$
  (c) $r = 8$      (d) $s = 3$

3 (a)

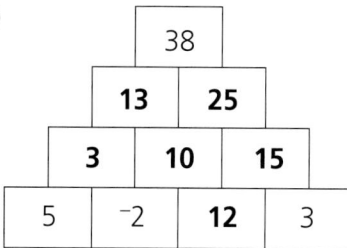

(b)

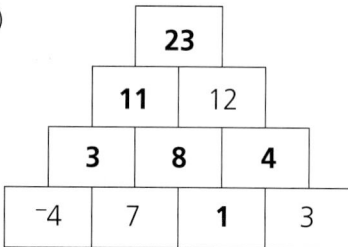

(c)

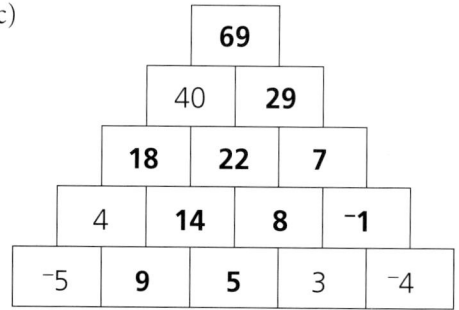

(d)

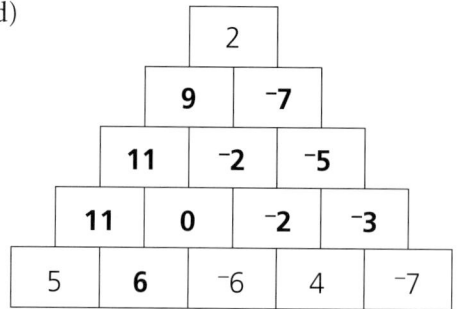

4 Both missing numbers are 4.

# ⑧ Fractions and decimals

---

**Practice booklet** pages 27 to 29

---

## Ⓐ Simplifying and comparing fractions (p 55)

## Ⓑ Adding and subtracting fractions (p 56)

## Ⓒ Multiplying a whole number by a fraction (p 56)

◊ '6 lots of $\frac{3}{4}$' and '$\frac{3}{4}$ of 6' do not 'feel' the same, even though they give the same result. You may need to spend some time dealing with different ways of understanding 'fraction × whole number' (or vice versa).

For example, pupils may find it helpful to see that $6 \times \frac{3}{4}$ is equivalent to $6 \times 3 \div 4$ and that $\frac{3}{4}$ of 6 is equivalent to $6 \div 4 \times 3$. As the order of '× 3' and '÷ 4' doesn't matter, the two expressions reduce to the same calculation.

The number line diagram shows that '6 lots of $\frac{3}{4}$' is 18 quarters, which make 4 whole ones and 2 quarters, or $4\frac{1}{2}$.

◊ This is a good point to introduce or revise cancelling common factors, because the fact that the result is as expected helps to convince pupils that the process is legitimate.

## Ⓓ Dividing a whole number by a fraction (p 57)

◊ Ordinary language is no help here. Someone who says 'Divide 10 by $\frac{1}{2}$' usually means 'Halve 10'.

You may need to give several whole-number examples to establish that A ÷ B can be thought of as 'How many Bs in A?'. For example, 15 ÷ 3 can mean 'How many 3s in 15?'.

◊ It will also help if you start with a number such as 12 and divide by smaller and smaller numbers. As the number you divide by gets smaller, the result gets larger. So pupils should expect $12 \div \frac{1}{2}$ to be greater than $12 \div 1$.

## E  Changing fractions to decimals (p 58)

◊ Some pupils may be fascinated by the idea of a never-ending process leading to a number which can never be completely written out (in decimal notation).

The mystery deepens when you ask them to multiply 0.3333… by 3. The result, as a decimal, is 0.9999… However, we know that $3 \times \frac{1}{3} = 1$, from which it follows that 0.9999… = 1.

Pupils may find the following argument helpful.

Let $\quad u = 0.9999…$
Then $\quad 10u = 9.9999…$
Subtract: $9u = 9$
So $\quad u = 1$

Some pupils may insist that 0.9999… is not equal to 1, that it differs from 1 by a tiny amount. Try asking how big this amount is.

### Investigation

The fractions which lead to terminating decimals are those whose denominators have only 2 and/or 5 as prime factors. In other words their denominators are of the form $2^m \times 5^n$ (where either $m$ or $n$ could be 0).

### Extension 1  Recurring decimals with a calculator

Pupils will have seen that the recurring digits in the decimals for $\frac{1}{7}$, $\frac{2}{7}$, … are cyclic. Although the pattern may be more complex in other cases, the cyclic property opens up a way of using a calculator to investigate recurring decimals.

For example, for $\frac{1}{17}$ a 10-digit calculator shows 0.058823529.

For $\frac{2}{17}$ it shows 0.117647058.

Because the group 058 occurs in both, it suggests that the decimal for $\frac{2}{17}$ continues like this: 0.117647058823529.

If we try $\frac{3}{17}$, we get 0.176470588, which does not help.

$\frac{4}{17}$ gives 0.235294117, which suggests the complete cycle in the decimal for $\frac{1}{17}$ is 0588235294117647.

(There may be a problem when the calculator automatically rounds up the last displayed digit if the next one is 5 or more.)

Sometimes there is no single cycle of digits but two or more separate cycles.

**Extension 2  How many recurring digits are there?**

The number of recurring figures is interesting but more difficult to generalise about. In the division process the recurrence starts when the same remainder appears twice. For example, when we divide 1 by 7, the remainders are 1, then 3, 2, 6, 4, 5, then 1 again:

$$\begin{array}{r} 0.142857 \\ 7\overline{)1.{}^10{}^30{}^20{}^60{}^40{}^50{}^10 \ldots} \end{array}$$

There are only six possible remainders (1, 2, 3, 4, 5, 6) and they are all used up in this process. If we are dividing by 13, there are 12 possible remainders; so the number of recurring figures cannot be more than 12. In fact, the process recurs after only six of the remainders are used. In the case of division by 17, all of the 16 remainders are used and there are thus 16 recurring figures.

## F  Mixed questions (p 59)

### A  Simplifying and comparing fractions (p 55)

**A1** (a) $\frac{4}{5} = \frac{\mathbf{16}}{20}$    (b) $\frac{12}{40} = \frac{\mathbf{3}}{10}$

(c) $\frac{3}{8} = \frac{\mathbf{9}}{24}$    (d) $\frac{2}{5} = \frac{10}{\mathbf{25}}$

(e) $\frac{16}{30} = \frac{\mathbf{48}}{90}$    (f) $\frac{2}{7} = \frac{10}{\mathbf{35}}$

(g) $\frac{8}{9} = \frac{32}{\mathbf{36}}$    (h) $\frac{4}{11} = \frac{32}{\mathbf{88}}$

(i) $\frac{12}{20} = \frac{\mathbf{3}}{5}$    (j) $\frac{1}{5} = \frac{\mathbf{5}}{25}$

**A2** (a) $\frac{9}{10}$ (b) $\frac{3}{7}$ (c) $\frac{5}{9}$ (d) $\frac{3}{8}$
(e) $\frac{3}{10}$ (f) $\frac{2}{5}$ (g) $\frac{1}{4}$ (h) $\frac{2}{9}$
(i) $\frac{3}{7}$ (j) $\frac{15}{28}$

**A3** (a) $\frac{11}{12}$ (b) $\frac{7}{9}$ (c) $\frac{5}{7}$ (d) $\frac{5}{12}$
(e) $\frac{5}{6}$ (f) $\frac{2}{3}$ (g) $\frac{3}{10}$ (h) $\frac{4}{5}$
(i) $\frac{2}{3}$ (j) $\frac{11}{16}$

**A4** (a) Albert has $\frac{9}{15}$ of a pizza, Bess has $\frac{10}{15}$ of a pizza. So Charlie has $\frac{11}{15}$ of a pizza (because $\frac{30}{15} - \frac{9}{15} - \frac{10}{15} = \frac{11}{15}$). Charlie has most.

(b) Albert has least.

### B  Adding and subtracting fractions (p 56)

**B1** (a) $\frac{1}{12}$ (b) $\frac{3}{4}$ (c) $\frac{2}{5}$ (d) $\frac{23}{40}$

(e) $\frac{7}{12}$ (f) $\frac{7}{30}$ (g) $\frac{13}{40}$ (h) $\frac{17}{20}$

(i) $\frac{5}{24}$ (j) $\frac{1}{6}$

**B2** (a) $\frac{37}{60}$ (b) $\frac{79}{120}$ (c) $\frac{11}{30}$ (d) $\frac{13}{60}$

**B3** $\frac{3}{4}$ and $\frac{7}{10}$, because these are the largest two fractions.
$(\frac{3}{4} = \frac{30}{40}, \frac{3}{5} = \frac{24}{40}, \frac{5}{8} = \frac{25}{40}, \frac{7}{10} = \frac{28}{40})$

**B4** (a) $\frac{11}{20}$   (b) $\frac{5}{24}$   (c) $\frac{11}{30}$   (d) $\frac{1}{12}$

**B5** (a) $\frac{17}{20}$   (b) $3\frac{17}{20}$

**B6** (a) $\frac{5}{12}$   (b) $2\frac{5}{12}$

**B7** (a) $4\frac{5}{6}$   (b) $2\frac{1}{2}$   (c) $5\frac{7}{8}$   (d) $\frac{7}{12}$
    (e) $\frac{9}{20}$

## C Multiplying a whole number by a fraction (p 56)

**C1** (a) 5   (b) $7\frac{1}{2}$   (c) 3   (d) $2\frac{1}{4}$
    (e) $\frac{3}{4}$   (f) $3\frac{1}{3}$   (g) $10\frac{1}{2}$   (h) $3\frac{1}{5}$
    (i) $12\frac{1}{2}$   (j) $7\frac{1}{2}$

**C2** (a) $2\frac{2}{3}$   (b) 6   (c) $7\frac{1}{2}$   (d) 10
    (e) $2\frac{1}{3}$   (f) 4   (g) $2\frac{1}{4}$   (h) $3\frac{1}{2}$
    (i) 6   (j) $5\frac{5}{6}$

## D Dividing a whole number by a fraction (p 57)

**D1** 6

**D2** 18

**D3** 20

**D4** 10

**D5** 12

**D6** 6

**D7** 24

**D8** 8

**D9** (a) 36     (b) 12     (c) 18
    (d) 9     (e) 80     (f) 16

**D10** (a) 30     (b) 15     (c) 10

**D11** (a) … divide by the numerator of the fraction.
    (b) $2\frac{2}{3}$; the pupil's diagram

**D12** (a) 32     (b) 2     (c) 8
    (d) 18     (e) $\frac{7}{8}$     (f) 56

## E Changing fractions to decimals (p 58)

**E1** 0.666…

**E2** (a) 0.1111…
    (b) $\frac{2}{9} = 0.2222\ldots$, $\frac{3}{9} = 0.3333\ldots$, up to $\frac{8}{9} = 0.8888\ldots$

**E3** (a) 0.16666…     (b) 0.083333…

**E4** $\frac{2}{7} = 0.285714285714\ldots$

$\frac{3}{7} = 0.428571428571\ldots$

$\frac{4}{7} = 0.571428571428\ldots$

$\frac{5}{7} = 0.714285714285\ldots$

$\frac{6}{7} = 0.857142857142\ldots$

The figures 2, 8, 5, 7, 1, 4 are always in the same order, but the starting figure is different.

**E5** (a) 0.076923076923…
    (b) 0.0588235294117647
        recurs

## Investigation

A fraction leads to a recurring decimal when the prime factors of its denominator include numbers other than 2, 5. These are recurring:

$\frac{1}{21}$   $\frac{1}{30}$   $\frac{1}{11}$   $\frac{1}{15}$   $\frac{1}{36}$   $\frac{1}{52}$

## F Mixed questions (p 59)

**F1** $13\frac{1}{4}$ hours

**F2** (a) $\frac{5}{12}$     (b) $\frac{1}{12}$     (c) $\frac{1}{6}$

**F3** $10\frac{2}{3}$ rugs

**F4** (a) $\frac{5}{6}$
    (b) (i) $\frac{7}{12}$     (ii) 0.583333…

## Egyptian fractions

Here is one way for each given fraction.

$\frac{5}{8} = \frac{1}{2} + \frac{1}{8}$     $\frac{7}{8} = \frac{1}{2} + \frac{1}{4} + \frac{1}{8}$     $\frac{7}{12} = \frac{1}{2} + \frac{1}{12}$

$\frac{9}{10} = \frac{1}{2} + \frac{1}{3} + \frac{1}{15}$     $\frac{17}{30} = \frac{1}{2} + \frac{1}{15}$

## What progress have you made? (p 58)

1 (a) $11\frac{1}{4}$      (b) $12\frac{1}{2}$

2 (a) 72      (b) 12

3 (a) 0.35      (b) 0.09090909…

   (c) 0.63636363…

## Practice booklet

### Sections A and B (p 27)

1 (a) $\frac{6}{9}, \frac{8}{12}$    $\frac{2}{4}, \frac{5}{10}$    $\frac{8}{10}, \frac{12}{15}$

   (b) $\frac{5}{15}, \frac{2}{6}$    $\frac{18}{24}, \frac{6}{8}$    $\frac{10}{12}, \frac{5}{6}$

   (c) $\frac{15}{40}, \frac{9}{24}$    $\frac{8}{32}, \frac{3}{12}$    $\frac{2}{6}, \frac{4}{12}$

   (d) $\frac{12}{30}, \frac{6}{15}$    $\frac{1}{6}, \frac{4}{24}$    $\frac{3}{12}, \frac{5}{20}$

2 (a) $\frac{240}{360} = \frac{48}{72} = \frac{6}{9} = \frac{2}{3}$

   (b) $\frac{360}{480} = \frac{90}{120} = \frac{30}{40} = \frac{3}{4}$

3 (a) $\frac{1}{4}$   (b) $\frac{15}{21}$   (c) $\frac{9}{10}$   (d) $\frac{5}{12}$

4 (a) $\frac{11}{12}$   (b) $\frac{11}{20}$   (c) $\frac{11}{15}$   (d) $\frac{7}{15}$

   (e) $\frac{5}{18}$   (f) $1\frac{17}{18}$   (g) $\frac{15}{16}$   (h) $\frac{19}{24}$

   (i) $6\frac{7}{24}$   (j) $8\frac{47}{80}$

5 (a) $\frac{3}{4}, \frac{7}{8}, \frac{15}{16}$; next terms $\frac{31}{32}, \frac{63}{64}$

6 (a) $\frac{1}{4}, \frac{3}{8}, \frac{5}{16}$; next terms $\frac{7}{32}, \frac{9}{64}$

   (b) $\frac{3}{6}, \frac{7}{12}, \frac{15}{24}$; next terms $\frac{31}{48}, \frac{63}{96}$

   (c) $\frac{1}{6}, \frac{3}{12}, \frac{5}{24}$; next terms $\frac{7}{48}, \frac{9}{96}$

### Sections C and D (p 28)

1 (a) 2   (b) $2\frac{1}{4}$   (c) $5\frac{1}{2}$   (d) $2\frac{2}{5}$

   (e) $1\frac{2}{3}$   (f) $5\frac{1}{3}$   (g) $5\frac{3}{5}$   (h) $4\frac{1}{2}$

   (i) 15   (j) $3\frac{3}{5}$   (k) $1\frac{5}{7}$   (l) $5\frac{5}{8}$

2 (a) 15   (b) 10   (c) 15   (d) 10

3 (a) 12   (b) 12   (c) 15   (d) 50

4 (a) $13\frac{1}{2}$ (b) 8   (c) 8   (d) 30

5 $4\frac{1}{2}$; the pupil's diagram

6 (a) $6\frac{3}{4}$   (b) 12   (c) 12   (d) $8\frac{1}{3}$

### Sections E and F (p 29)

1 (a) 0.06666… and working

   (b) 0.03333…

2 (a) 0.28      (b) 0.83333…

   (c) 0.375     (d) 0.77777…

   (e) 0.025     (f) 0.27272…

3 (a) $\frac{1}{6}$      (b) 4 hours

4 (a) $6\frac{3}{4}$ hours    (b) $\frac{1}{2}$ hour more

5 (a) (i) $\frac{2}{3}$      (ii) 0.66666…

   (b) (i) $\frac{8}{15}$     (ii) 0.53333…

# Review 1 (p 60)

**1**

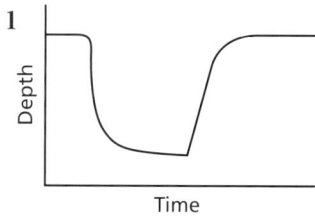

**2** (a) 2　(b) $\frac{1}{2}$　(c) $\frac{1}{2}$　(d) 4
　(e) $\frac{1}{3}$　(f) $1\frac{1}{2}$　(g) $\frac{1}{4}$　(h) $\frac{3}{4}$

**3** (a) $x = 5$　(b) $y = 5$　(c) $z = 11$

**4** (a)

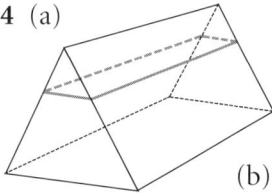

(b)

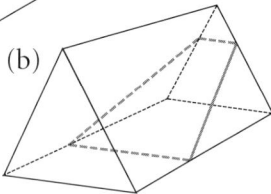

(c)

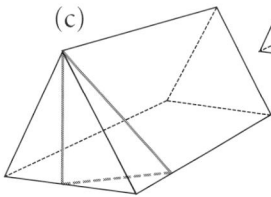

**5** (a) July　(b) Cannot tell
　(c) 16°C

**6** (a) (i) $75\,000\,\text{cm}^2$　(ii) $7.5\,\text{m}^2$
　(b) 1.5 litres

**7** (a) $a + 5$　(b) $3b + 3$　(c) $2 - 4c$
　(d) $6 - 3d$　(e) $7 - 3e$　(f) $f - 3$

**8** (a) 5　(b) 4

**9** (a) 16
　(b) A　$12 \div \frac{3}{4}$

**10** (a) $x = 6$　(b) $y = {}^-3$　(c) $z = 5$
　(d) $x = 6$　(e) $x = 8\frac{1}{2}$　(f) $x = {}^-4$

**11** (a) 30 km　(b) 1 cm to 6 km
　(c) 1 cm to 4 km

**12** 0.307 692 307 692 …

**13** $^-1$

**14** (a) 0.03333…
　(b) 0.36666…
　(c) $\frac{11}{30}$
　(d) $0.3 = \frac{3}{10}$
　(e) $\frac{1}{3} - \frac{1}{30} = \frac{10}{30} - \frac{1}{30} = \frac{9}{30} = \frac{3}{10}$

## Mixed questions 1 (Practice booklet pp30)

**1** (a) 0.75 km　(b) $0.03375\,\text{km}^2$

**2** (a) $x = \frac{1}{2}$　(b) $y = {}^-5$　(c) $z = 7$
　(d) $a = 11$　(e) $b = 20$　(f) $c = 5$

**3** (a)
$$
\begin{array}{r}
0.1\,6\,6\,6\,\ldots \\
6\,\overline{)\,1\,.0^4 0^4 0^4 0\,\ldots}
\end{array}
$$
　(b)
$$
\begin{array}{r}
0.1\,6\,6\,6\,\ldots \\
-\,0.1\,0\,0\,0\,\ldots \\
\hline
0.0\,6\,6\,6\,\ldots
\end{array}
$$
　(c) $\frac{1}{15}$

**4** (a) $4x - 7$　(b) $4y - 7b - 8$
　(c) $5 - 3z$

**5** (a) ▢　▢
　(b) ▭
　(c) ▭

**6** 30

# 9 Transformations

| **Essential** | **Optional** |
|---|---|
| Sheets 193 to 197 | OHP transparencies of pages 63 and 65 |
| Tracing paper | Small mirrors |
| Spreadsheet (section G) | |
| **Practice booklet** pages 31 to 35 | |

## A Reflection (p 63)

> Sheet 193

◊ If you use an OHP transparency of the diagram, pupils can indicate the mirror lines on it.

   If not, they will need to describe where the lines are. They may use equations (and you may wish to do section B first), or they can describe lines as, for example, 'the line parallel to the $x$-axis through $(0, 6)$', 'the line through $(2, 6)$ and $(7, 6)$'. Encourage them to describe the lines precisely: they may see how much more economical it is to use equations.

◊ If mirrors or tracing paper are used, they should be just for checking.

## B Lines on a grid (p 64)

◊ Pupils often confuse lines of the form $x = a$ and those of the form $y = a$. It is helpful to list some of the points on a line such as $x = 2$ and note that the $x$-coordinate of every point on the line is 2.

## C Rotation (p 65)

> Tracing paper, sheets 194 to 196

◊ Tracing paper is likely to be needed when the centre of rotation is not on the line of one of the edges of the shape. Even so, encourage pupils to try to find centres without tracing paper and then to check with it afterwards.

◊ A very useful technique when rotating a shape is to connect the centre with a point on the shape by an 'L' whose lines are parallel to the axes. The 'L' is then easy to rotate.

◊ The rotations in the diagram were done as follows:

to A: 180° about $(11, 5)$    to B: 90° anticlockwise about $(8, 6)$

to C: 90° clockwise about $(12, 6)$    to D: 180° about $(9\frac{1}{2}, 6\frac{1}{2})$

to E: 90° anticlockwise about $(4, 6)$

## D Translation (p 66)

## E Enlargement from a centre (p 67)

> Sheet 197

◊ Emphasise that all distances must be measured from the centre of enlargement. (It is a common error to measure the enlarged distance from the point on the object rather than from the centre.)

## F Describing transformations (p 68)

## G Using a spreadsheet (p 70)

◊ It is possible to use a program specifically designed for transformations. However, these programs do not usually allow the pupil to see the arithmetical effect on the coordinates of individual points.

◊ With some spreadsheets, changing axes to have a central origin may require a little forethought, or this may be left alone.

Leaving a gap between the sets of coordinates is essential if the shapes are to be drawn correctly.

## H Combining transformations (p 72)

◊ The Escher picture on page 73 can be used for class discussion.

## A Reflection (p 63)

**A1** The pupil's completed sheet 193

## B Lines on a grid (p 64)

**B1** (a) $x = {}^-2$  (b) $y = {}^-3$

**B2** (a) Because at every point on the line the $y$-coordinate is equal to the $x$-coordinate

(b) Because at every point the $y$-coordinate is the negative of the $x$-coordinate

**B3** (a) The $y$-axis  (b) The $x$-axis

**B4** (a) $(1, 2)$  (b) $(3, 0)$

(c) $(2, 3)$  (d) $({}^-2, {}^-3)$

**B5** $(1, {}^-4)$

**B6** $({}^-4, {}^-3)$

## C Rotation (p 65)

**C1** The pupil's completed sheet 195

**C2** (a) $(5, 4)$

(b) Add the $x$-coordinates of A and U and divide by 2. Do the same with the $y$-coordinates.

(c) B is $(1, 3)$. V is $(9, 5)$.
Midpoint of BV is
$$\left(\frac{1 + 9}{2}, \frac{3 + 5}{2}\right) \text{ i.e. } (5, 4)$$

(d) Midpoint of CW is
$$\left(\frac{2 + 8}{2}, \frac{1 + 7}{2}\right) \text{ i.e. } (5, 4)$$

Midpoint of DX is
$$\left(\frac{11 + {}^-1}{2}, \frac{{}^-2 + 10}{2}\right) \text{ i.e. } (5, 4)$$

Midpoint of EY is
$$\left(\frac{{}^-2 + 12}{2}, \frac{{}^-4 + 12}{2}\right) \text{ i.e. } (5, 4)$$

Midpoint of FZ is
$$\left(\frac{{}^-3 + 13}{2}, \frac{{}^-1 + 9}{2}\right) \text{ i.e. } (5, 4)$$

(e) M is the centre for the rotation of 180° which moves ABCDEF on to UVWXYZ.

## D Translation (p 66)

**D1** (a) $\begin{bmatrix} 2 \\ 1 \end{bmatrix}$ A, $\begin{bmatrix} 4 \\ -2 \end{bmatrix}$ D, $\begin{bmatrix} -4 \\ -2 \end{bmatrix}$ C

(b) B $\begin{bmatrix} -2 \\ 1 \end{bmatrix}$

**D2** Lauren is wrong. The column vector is $\begin{bmatrix} 4 \\ 1 \end{bmatrix}$.

**D3** (a), (b)

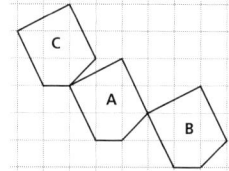

(c) (ii) $\begin{bmatrix} -5 \\ 3 \end{bmatrix}$  (ii) $\begin{bmatrix} 5 \\ -3 \end{bmatrix}$

**D4** Translation with column vector $\begin{bmatrix} -3 \\ -1 \end{bmatrix}$

## E Enlargement from a centre (p 67)

**E1** The pupil's completed sheet 197

**E2** P goes to $(4, 6)$.
Q goes to $(8, 4)$.
R goes to $(6, {}^-4)$.

**E3**

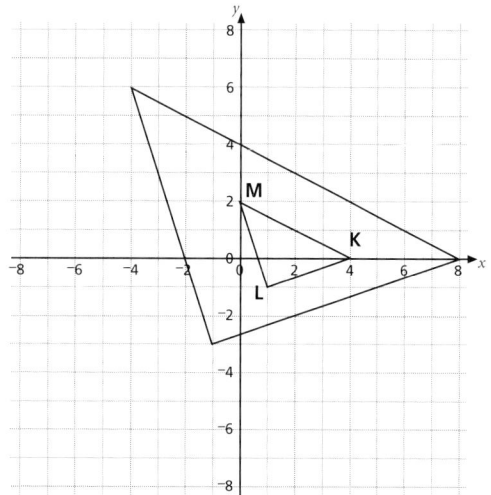

$(8, 0)$ $({}^-1, {}^-3)$ $({}^-4, 6)$

## F Describing transformations (p 68)

**F1** (a) Reflection in the $x$-axis

(b) Translation $\begin{bmatrix} 6 \\ -5 \end{bmatrix}$

(c) Reflection in the $y$-axis

(d) Reflection in $y = 6$

(e) Rotation 90° anticlockwise about (0, 0)

(f) Rotation 180° about (0, 0)

(g) Reflection in $y = x$

(h) Translation $\begin{bmatrix} -6 \\ 5 \end{bmatrix}$

(i) Reflection in $x = 4$

(j) Enlargement, centre ($^-$1, $^-$5), scale factor 2

**F2** B: Rotation 180° about O

C: Rotation 90° anticlockwise about O

D: Reflection in line $q$

E: Reflection in line $r$

**F3** A: Rotation 180° about (6, 5)

B: Translation $\begin{bmatrix} 6 \\ 1 \end{bmatrix}$

C: Rotation 180° about $(7, 3\frac{1}{2})$

D: Translation $\begin{bmatrix} -4 \\ 1 \end{bmatrix}$

E: Rotation 180° about $(6, 2\frac{1}{2})$

**F4** (a) A: Translation $\begin{bmatrix} 1 \\ 3 \end{bmatrix}$

B: Rotation 90° clockwise about (4, 4)

C: Reflection in $y = 5$

D: Rotation 180° about (7, 3)

*(b) E: Rotation 90° clockwise about $(4\frac{1}{2}, 5\frac{1}{2})$

Ⓖ **Using a spreadsheet** (p 70)

**G1** (a) (i) Add 2 to each $x$-coordinate.

(ii) Add 3 to each $x$-coordinate and 5 to each $y$-coordinate.

(b) (i) Subtract 5 from each $x$-coordinate.

(ii) Take 2 from the $x$-coordinates and 4 from the $y$-coordinates.

**G2** (a) $x$ stays the same. $y$ is multiplied by $^-$1.

(b) $x$ and $y$ swap over.

**G3** (a) Multiply $x$ and $y$ by $^-$1.

(b) Swap over $x$ and $y$ and multiply the new $x$-coordinate by $^-$1.

(c) Swap over $x$ and $y$ and multiply the new $y$-coordinate by $^-$1.

**G4** (a) It enlarges the shape by a scale factor of 2, with centre (0, 0).

(b) Shapes are enlarged by a scale factor of the whole number, with centre (0, 0).

(c) (i) Enlargement scale factor $\frac{1}{2}$

(ii) Enlargement scale factor $\frac{1}{4}$

(d) The tranformation is an enlargement with scale factor $^-$2, but as a negative scale factor has not been introduced so far, the transformation could be described as enlargement × 2 followed by 180° rotation about (0, 0), or vice versa.

*$\star$**G5** (a) $x$ changes to $(8 - x)$, $y$ stays the same.

(b) $x$ changes to $(4 - x)$, $y$ changes to $(4 - y)$.

Ⓗ **Combining transformations** (p 72)

**H1** (a) T

(b) Rotation 180° about (0, 0)

**H2** (a) W

(b) Reflection in the $y$-axis

(c) U

(d) Reflection $y = x$

(e) S

(f) Rotation 90° clockwise about (0, 0)

(g) Reflection in $y = x$

(h) Rotation 90° clockwise about (0, 0)

(i) Reflection in the $y$-axis

(j) Reflection in $y = x$

(k) Reflection in the $y$-axis

(l) T

(m) R

(n) Reflection in the $x$-axis

(o) P

(p) Reflection in $y = ^-x$

(q) W

(r) Reflection in the $x$-axis

## What progress have you made? (p 73)

**1**

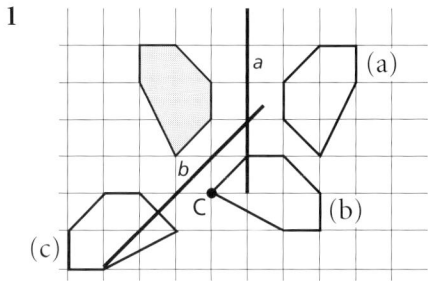

**2** (a) Reflection in $y = 3$

(b) Rotation 90° anticlockwise about $(3, 3)$

(c) Reflection in $y = x$

(d) Rotation 180° about $(6, 4\frac{1}{2})$

(e) Translation $\begin{bmatrix} -2 \\ -3 \end{bmatrix}$

(f) Rotation 180° about $(5, 3)$

**3** Reflection in $y = x$

## Practice booklet

### Sections A and B (p 31)

**1**

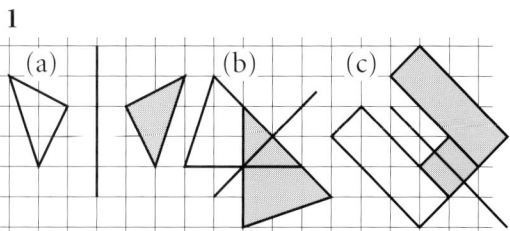

**2** (a)

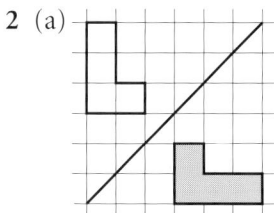

(b)

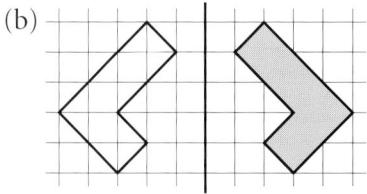

**3** (a) $y = x$    (b) $y = 3$

## Sections C and D (p 32)

**1** (a)

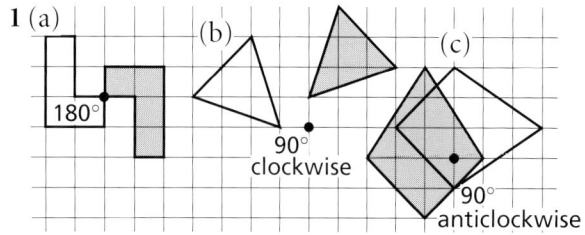

**2** A: Rotation 180° about $(4, 3)$

B: Rotation 90° anticlockwise about $(12, 2)$

**3** (a) $(0, 1)$    (b) $(2, 3)$

**4** (a) Translation by vector $\begin{bmatrix} 1 \\ -2 \end{bmatrix}$

(b) Translation by vector $\begin{bmatrix} -1 \\ 2 \end{bmatrix}$

## Section E (p 33)

**1**

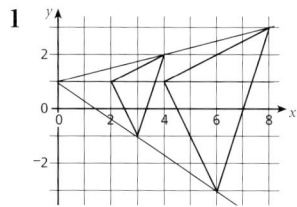

Vertices of the enlargement
$(4, 1), (8, 3), (6, {}^-3)$

**2**

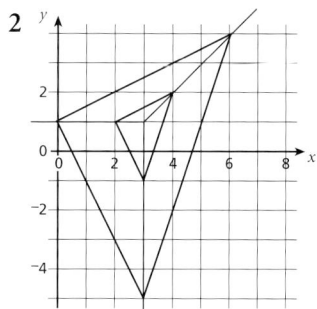

Vertices of the enlargement
$(0, 1), (6, 4), (3, {}^-5)$

## Section F (p 34)

**1** (a) Reflection in $y = 0$ ($x$-axis)

(b) Reflection in $y = {}^-3$

(c) Translation by vector $\begin{bmatrix} 0 \\ 6 \end{bmatrix}$

(d) Rotation of 180° about $(0, 3)$

(e) 180° rotation about $(0, 0)$

(f) Rotation of 90° clockwise about $(0, 0)$

(g) Rotation of 90° clockwise about $(0, 0)$

(h) Rotation of 180° about $(0, 0)$

(i) Reflection in $y = 0$ ($x$-axis) or Translation by vector $\begin{bmatrix} 0 \\ 6 \end{bmatrix}$

(j) Reflection in $y = {}^-x$

(k) Reflection in $y = 3$

(l) Rotation of 90° anticlockwise about $(0, 0)$

(m) Rotation of 180° about $(0, 0)$

(n) Reflection in the line $y = {}^-x$

**Section H** (p 35)

1 (a)

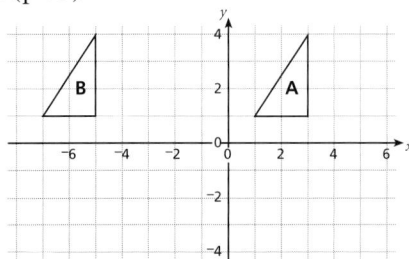

(b) Translation by vector $\begin{bmatrix} {}^-8 \\ 0 \end{bmatrix}$

2 (a)

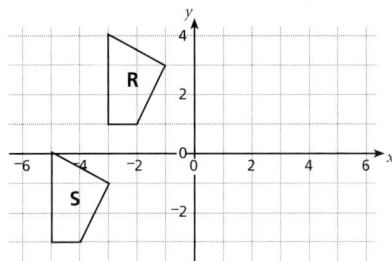

(b) Translation by vector $\begin{bmatrix} {}^-2 \\ {}^-4 \end{bmatrix}$

# ⑩ True, iffy, false 1

These questions are intended to promote discussion. You could ask pupils to spend a short while on them individually before coming together in pairs or small groups to exchange ideas. Then you could take a vote on whether each question is true, iffy or false before discussing them with the whole class.

1 Iffy
   To multiply a whole number by 10, add an extra 0 at the end of the number.

2 True

3 True

4 Iffy
   Squaring a number greater than 1 or less than 0 gives a higher number.

5 False

6 Iffy
   The total of two positive numbers is greater than their difference.

7 True

8 Iffy
   A prime number greater than 2 is an odd number.

9 True

10 False

# 11 Linear equations and graphs

| **Essential** | **Optional** |
|---|---|
| Sheets 198 and 199 | Graph plotting computer program |
| **Practice booklet** pages 36 to 38 | |

## Ⓐ Gradient (p 75)

◊ You can use the lines in the pupil's book to introduce the question 'How can we measure the steepness of a line?'

Some pupils may suggest using angles. Others may refer to road signs, where the steepness of a hill may be given as a percentage, or in the form '1 in 6'.

The latter is sometimes used to mean '1 up for every 6 horizontally' and sometimes '1 up for every 6 along the road'. The first interpretation is more useful here. (With the gradients found on roads the difference between the two interpretations is negligible in practice.)

◊ When the idea of gradient as the ratio up/across has been introduced, emphasise that whichever two points are chosen the calculated gradient will be the same.

◊ Ask pupils how they would distinguish between the gradients of upward and downward sloping lines.

◊ The gradients of the given lines are:    *a* 2    *b* 1.5    *c* 0.75    *d* ⁻0.4

## Ⓑ Investigating linear graphs (p 77)

The aim is for pupils to see that for the graph of $y = mx + c$, $m$ controls the gradient and $c$ the intercept on the $y$-axis.

Optional: graph plotting computer program

◊ Using a graph plotting program speeds up the investigation, allowing pupils to experiment with a large number of different equations.

Show pupils how to place the axes roughly as shown in the diagram, so that all four quadrants are shown. The axes must be scaled equally.

◊ Although a spreadsheet (with graphing facility) or graphic calculator may appear to be a suitable substitute for a graph plotting program, each has problems. It is difficult to control the scaling of the axes when using a spreadsheet. On a graphic calculator the low resolution makes it difficult to read off gradients, although the general effect of changing $m$ and $c$ can be observed.

**Extension** Pupils often find it challenging to use the graph plotting program to draw geometrical shapes. Such problems are stimulating and help reinforce the significance of $m$ and $c$.

Examples that teachers have used include:

- Make a square with four lines.
  Now make a square twice as big.

- Make a square that has one corner at (1, 1).

- Make a rectangle.
  Now make a rectangle that is twice as long as it is wide.

## C Gradient and intercept (p 78)

## D Negative gradients (p 79)

Sheets 198 and 199

### A Gradient (p 75)

**A1** (a) 0.11 Rack and pinion (train)
(b) 0.43 Rack and pinion (single car)
(c) 0.67 Funicular
(d) 0.25 Rack and pinion (single car)
(e) 0.12 Rack and pinion (train)

### C Gradient and intercept (p 78)

**C1** (a) (i) 7 (ii) 2
(b) (i) 0.2 (ii) ⁻4
(c) (i) 8 (ii) 0
(d) (i) 1 (ii) ⁻7
(e) (i) 0 (ii) 5

**C2** (a) $y = 4x + 7$ (b) $y = \frac{1}{2}x + 9$
(c) $y = x + 4$ (d) $y = \frac{1}{3}x + 2$
(e) $y = 2x - 6$ (f) $y = \frac{1}{2}x - 3$
(g) $y = \frac{1}{2}x - 9$

### D Negative gradients (p 79)

**D1** (a) (i) ⁻4 (ii) 1
(b) (i) $-\frac{1}{2}$ (ii) ⁻6
(c) (i) ⁻1.5 (ii) 0
(d) (i) ⁻2 (ii) 6
(e) (i) ⁻1 (ii) 7

**D2** (a) $y = {}^-2x + 10$ or $y = 10 - 2x$

(b) $y = {}^-3x + 6$ or $y = 6 - 3x$

(c) $y = -\frac{1}{2}x + 4$ or $y = 4 - \frac{1}{2}x$

(d) $y = {}^-2x - 5$

(e) $y = -\frac{1}{2}x - 3$

**D3** (a) $y = x - 2$,  $y = x$,
$y = x + 3$,  $y = x + 5$

(b) $y = 2x - 3$,  $y = 2x$,  $y = 2x + 4$

(c) $y = x$,  $y = 3x + 2$
$y = {}^-x + 8$

(d) $y = \frac{1}{2}x$,  $y = 2x$,
$y = 4x$,  $y = {}^-4x$

(e) $y = x + 3$,  $y = {}^-x$,
$y = {}^-x + 3$,  $y = {}^-x - 4$

(f) $y = x$,  $y = {}^-x$,  $y = x^2$

**D4** (a)

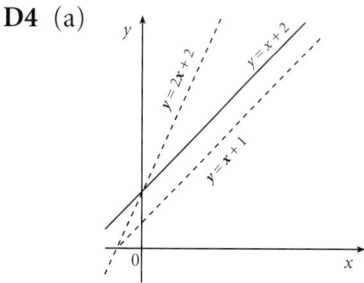

(b)

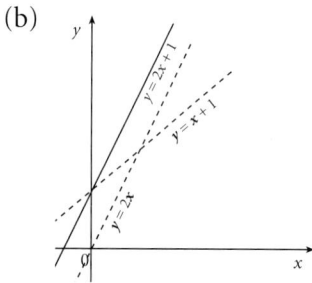

(c)

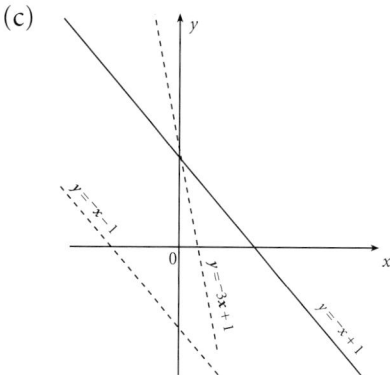

(d)

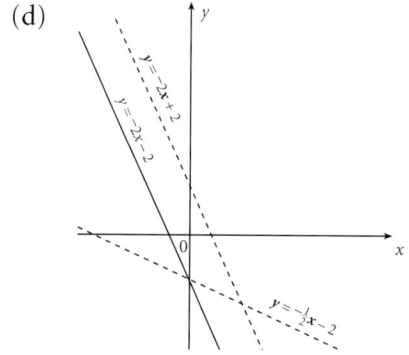

(e)

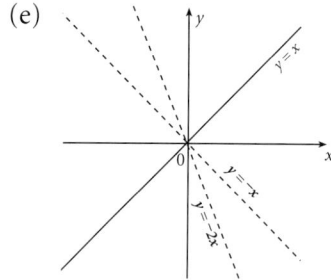

**D5** A is $y = {}^-x + 2$ or $y = 2 - x$
B is $y = x - 2$ or $y = x + {}^-2$

**D6** A is $y = 2x + 1$,   B is $y = 2x - 2$,
C is $y = 1 - \frac{1}{2}x$,   D is $y = -\frac{1}{2}x - 2$

## What progress have you made? (p 80)

**1** (a)  1.5      (b)  $-\frac{1}{2}$

**2** (a)  (i)  3      (ii)  $^-4$
(b)  (i)  $^-2$      (ii)  4

**3** (a)  $y = \frac{1}{2}x - 1$
(b)  $y = {}^-2x + 2$  or  $y = 2 - 2x$

## Practice booklet

## Section A (p 36)

**1** (a)  0.2      (b)  0.4      (c)  1.33
(d)  0.33      (e)  0.125

**2** (a)  0.05, single carriageway
(b)  0.04, dual carriageway
(c)  0.06, single carriageway

## Section C (p 37)

1. (a) (i) 4     (ii) 3
   (b) (i) $\frac{1}{2}$     (ii) 0
   (c) (i) 3     (ii) $^-1$
   (d) (i) 0     (ii) 4
   (e) (i) 0.1     (ii) $^-3$
   (f) (i) 7     (ii) 6

2. (a) $x = 6$     (b) $y = x + 3$
   (c) $y = 2x - 8$     (d) $y = \frac{x}{4} + 8$
   (e) $y = \frac{x}{5}$     (f) $y = {}^-4$

## Section D (p 38)

1. (a) (i) $^-3$     (ii) 0
   (b) (i) $^-1$     (ii) 4
   (c) (i) 2     (ii) 5
   (d) (i) $^-0.2$     (ii) $^-3$
   (e) (i) $^-4$     (ii) 3
   (f) (i) 2     (ii) $-\frac{1}{2}$

2. A is $y = x + 2$
   B is $y = x$
   C is $y = 2x$
   D is $y = 2 - x$

3
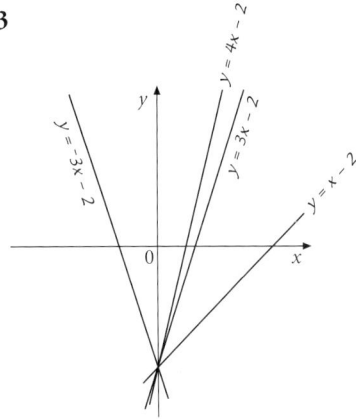

4
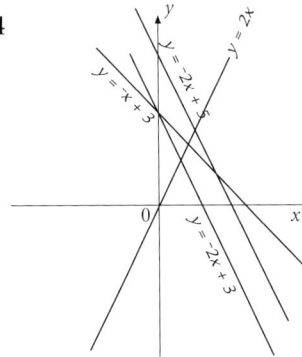

# 12 Percentage changes

**Practice booklet** pages 39 to 43

## A Percentages and decimals (p 81)

You can use the scale in the pupil's book to introduce the decimal equivalents of percentages like 64.5%.

## B Percentage increases (p 82)

◊ The multiplier approach is used here, as it easily extends to successive percentage changes (for example, compound interest).

◊ To get pupils thinking about the matter you can pose a question before they see the method in the book: 'I want to increase something by 50%, but I want to do it in one step (or with only one operation). What should I do? What if I want to increase something by 25%, and so on?'

◊ If pupils are able to see for themselves how to use a single multiplier for percentage increases, you can ask them to do the same for a percentage decrease, say 30%.

## C Percentage decreases (p 83)

The multiplier approach is extended to percentage decreases.

## D Successive increases and decreases (p 84)

**D6** It is easier to see the result if you give a value to the original number of fish, say 100.

## E Expressing changes as percentages (p 85)

Pupils need to be secure in understanding that to find the multiplier from A to B you have to divide B by A.

## F Problems (p 86)

**\*F5** In case pupils think that chocolate really was rationed in 1984, it should be pointed out that '1984' here refers to the novel by George Orwell.

## A Percentages and decimals (p 81)

**A1** (a) 0.3     (b) 0.32     (c) 0.325
      (d) 0.396    (e) 0.403

**A2** (a) 70%     (b) 74%     (c) 74.8%
      (d) 40.6%    (e) 33.3%

**A3** (a) 0.06     (b) 0.063    (c) 0.045
      (d) 0.019    (e) 0.007

**A4** (a) 2%      (b) 2.6%     (c) 9.1%
      (d) 0.5%     (e) 15.75%

**A5** (a) £84.00    (b) £84.72    (c) £303.75
      (d) £27.20    (e) £126.63    (f) £49.50
      (g) £31.32    (h) £5.28

**A6** (a) 0.125       (b) 12.5%

**A7** (a) 0.625       (b) 62.5%

**A8** 68.75%

**A9** 59.6%

## B Percentage increases (p 82)

**B1** (a) £50      (b) £5      (c) £75
      (d) £52.50    (e) £2.25

**B2** 1.15

**B3** (a) £34.50    (b) £9.20    (c) £57.50
      (d) £87.40    (e) £1.38

**B4** (a) £56      (b) £95.20    (c) £7.28

**B5** Karl is wrong. You multiply by 1.08.

**B6** (a) £54      (b) £7.56    (c) £21.60
      (d) £38.88    (e) £1.62

**B7** (a) 1.045       (b) £18.81

**B8** (a) £45.54    (b) £70.86    (c) £121.08
      (d) £42.56

## C Percentage decreases (p 83)

**C1** (a) £17      (b) £32.30    (c) £61.20
      (d) £58.14    (e) £42.16

**C2** 0.65

**C3** (a) £19.50    (b) £49.40    (c) £117
      (d) £64.09    (e) £160.68

**C4** 0.94

**C5** (a) £41.36    (b) £26.32    (c) £573.40
      (d) £357.20    (e) £1381.80

**C6** (a) 0.97    (b) 0.91    (c) 0.89
    (d) 0.7    (e) 0.855

**C7** (a) £220.80    (b) £720    (c) £53.40
    (d) £69.12

**C8** (a) £47.25    (b) £57.20    (c) £100.58
    (d) £233.80

**C9** (a) 17% increase  (b) 2% increase
    (c) 6% decrease    (d) 12% decrease
    (e) 40% increase    (f) 30% decrease
    (g) 72.5% increase
    (h) 35% decrease  (i) 76.5% decrease

### D Successive increases and decreases (p 84)

**D1** (a) 23 000    (b) 24 840

**D2** (a) £6468    (b) £8316
    (c) £12 127.50    (d) £14 495.25
    (e) £20 790

**D3** It makes no difference. Multiplying first by 1.05 and then by 1.1 gives the same result as multiplying first by 1.1 and then by 1.05.

**D4** (a) 0.95
    (b) (i) £68.40    (ii) £62.70
        (iii) £10.03    (iv) £21.09
        (v) £461.70
    (c) It makes no difference.

**D5** (a) £119.60  (b) £106.40  (c) £477.05
    (d) £8.88    (e) £8.85

**D6** Karl is wrong. The population is multiplied by 1.1 and then by 1.1 again. So overall it is multiplied by $1.1 \times 1.1$, which is 1.21. This is a 21% increase.

### E Expressing changes as percentages (p 85)

**E1** 8%

**E2** (a) 14%    (b) 24%    (c) 15%

    (d) 34%    (e) 12.5%    (f) 11.8%

**E3** 12%

**E4** 5.2%

**E5** (a) 5%    (b) 15%    (c) 19%
    (d) 23%    (e) 45%    (f) 2%

**E6** (a) 8% increase    (b) 14% decrease
    (c) 7% decrease    (d) 34% decrease
    (e) 7.5% increase  (f) 11.5% decrease

### F Problems (p 86)

**F1** (a) 25%        (b) 20%

**F2** £97.24

**F3** (a) 40% (b) 82% (c) £16 (d) 30%

**\*F4** 44%

**\*F5** 1% decrease

**\*F6** 5 years

### What progress have you made? (p 86)

**1** (a) £372.60    (b) £141.90

**2** (a) 16% increase  (b) 7% decrease

**3** £239

## Practice booklet

### Section A (p 39)

**1** (a) 0.55    (b) 0.6
   (c) 0.655    (d) 0.846

**2** (a) 39%    (b) 40%
   (c) 42.5%    (d) 50.9%

**3** (a) 0.04    (b) 0.042
   (c) 0.065    (d) 0.008

**4** (a) 3%    (b) 3.5%
   (c) 4.8%    (d) 18.25%

**5** 8%, 0.08, 8 out of 100
   80%, 8 out of 10, 80 out of 100, 0.8
   0.8%, 0.008, 8 out of 1000

6  (a)  £33.60        (b)  £18.75
   (c)  £35.12        (d)  £3.12
   (e)  £3.23         (f)  £0.29

7  (a)  0.225         (b)  22.5%

8  67.5%

9  69.6%

10  1.5%,  3 out of 200,  0.015,
    $1\frac{1}{2}$ out of 100,  15 out of 1000

    15%,  15 out of 100,  12 out of 80,
    3 out of 20,  0.15

## Sections B and C (p 40)

1  (a)  £28          (b)  £44.80
   (c)  £20.16       (d)  £123.20

2  (a)  £4.20        (b)  £2.10
   (c)  £29.40       (d)  £12.18

3  1.125

4  (a)  £4.50        (b)  £2.70
   (c)  £13.50       (d)  £4.05

5  0.925

6  (a)  £37          (b)  £8.14
   (c)  £14.80       (d)  £333

7  (a)  1.17         (b)  1.265
   (c)  1.07         (d)  1.036

8  (a)  0.82         (b)  0.97
   (c)  0.895        (d)  0.952

9  (a)  35% increase  (b)  60% increase
   (c)  25% decrease  (d)  9% decrease
   (e)  70% decrease  (f)  16.5% increase
   (g)  4.5% decrease  (h)  2.5% increase

## Section D (p 41)

1  (a)  £8.40        (b)  £9.24

2  (a)  £137.70      (b)  £267.75
   (c)  £55.08       (d)  £9.18

3  (a)  £4984.32     (b)  £13 083.84
   (c)  £6490        (d)  £25 960

4  (a)  £275.23      (b)  £45.56
   (c)  £1750.27     (d)  £48 428.80

5  (a)  Lower.  $1.1 \times 0.9 = 0.99$
   (b)  Yes, the same.  $0.9 \times 1.1 = 1.1 \times 0.9$

## Section E (p 42)

1  (a)  1.12         (b)  12%

2  (a)  0.92         (b)  8%

3  (a)  4% increase   (b)  12% increase
   (c)  4% decrease   (d)  27% decrease
   (e)  24% increase  (f)  9% decrease

4  (a)  12.5% decrease
   (b)  3.5% increase
   (c)  7.2% increase
   (d)  21.8% decrease

5  (a)  7.8%          (b)  383, 353

## Section F (p 43)

1  Tony 8.5%,  John 9.0%.  John

2  (a)  £52.25        (b)  £62.31

3  (a)  6 078 000 000  (b)  1.3%

4  No.  $0.75 \times 0.75 = 0.5625$

5  $0.8 \times 0.8 = 0.64$
   The population will have gone down by
   36%. His estimate will be too low.

6  Between 6 and 7 years

# ⑬ True, iffy, false 2

◊ These questions are intended to promote discussion. You could ask pupils to spend a short while on them individually before coming together in pairs or small groups to exchange ideas. Then you could take a vote on whether each question is true, iffy or false before discussing them with the whole class.

**1** False

**2** Iffy.
The diagonals of a rectangle cross at right angles if the rectangle is a square.

**3** Iffy.
A parallelogram has no lines of symmetry unless it is a rhombus.

**4** Iffy.
Four straight lines cross at six points unless any of them are parallel to each other.

**5** True

**6** Iffy.
A right-angled triangle has one line of symmetry if it is isosceles.

**7** Iffy.
If B, C and A are on a straight line in that order, then if A is 20 cm from B and B is 10 cm away from B then A and C are 10 cm apart.

**8** True

**9** Iffy.
If two rectangles overlap, the shape of the overlap is also a rectangle if each pair of sides of one rectangle is parallel to a pair of sides of the other.

# ⑭ Probability from experiments

| **Essential** | **Optional** |
|---|---|
| Plastic spoons (one per pair) | OHP transparency of sheet 200 |
| Sheet 200 | Multilink cubes (10 per group, see below) |
| Dice | Drawing pins (one per group, see below) |
| | Sheet 201 |
| **Practice booklet** pages 44 and 45 | |

## A Experiments (p 88)

> Plastic spoons (one per pair), sheet 200
> Optional: Multilink cubes (10 per pair doing the experiment), drawing pins (one per pair), sheet 201 (one per pair)

◊ You could start by holding up a plastic spoon and asking what will happen if it is dropped. Make it clear that although we might have a guess at the probability of it landing the right way up, the spoon is not symmetrical, so we can't use 'equally likely outcomes'.
The 0 to 1 scale is a useful way of thinking about probability. You could mark some of the class's guesses on a large scale on the board.
In one school the teacher had half the class doing the spoon experiment and the other half the drawing pin experiment.

◊ When pupils (in pairs) do their own trials, keep the recording informal for the first set of trials and collect together the class's results. (The complexity of the resource sheet could be counter-productive at first.)
When using the sheet, explain the first 'block' (recording the outcomes) first. When this has been completed, go on to explain the next block of columns, then the relative frequency, and finally the graph.
The graph should show how the relative frequency 'settles down' to a value which can be used as an estimate of the probability.

◊ If everybody has used identical spoons, the class's results can be pooled to get a better estimate.

### Dropping a multilink cube

The design of a suitable data collecting sheet can be left to the pupils.

### Drawing pin experiment

Some pupils could do a comparative experiment, for example comparing two different surfaces.

### Money down the drain

You can discuss whether this is entirely a matter of chance.

If it is, then the probability that the penny lands entirely in a white area is actually quite small. The diagram below explains why.

The left hand edge of the penny can finish anywhere along a line of this length.

But the penny goes down the drain only if its left hand edge finishes in one of these sections of the line.

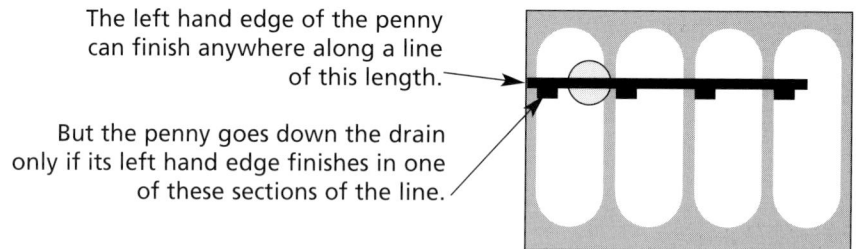

## B **Relative frequency** (p 91)

◊ You may need an introductory session on changing fractions to decimals.

**B4** The relative frequencies add up to 1.01, because of rounding errors.

## C **Estimating probabilities** (p 93)

*'Lovely discussions around the room on games and chance. All designed their own game; these were displayed in the room.'*

| Dice |
| --- |

**C1** The probability of winning can be worked out theoretically (but not by pupils at this level!). It is $\frac{3}{7}$, or approximately 0.43.

**C2** The theoretical probability that A wins is $\frac{3}{8}$. Adding an extra square increases this to $\frac{17}{32}$.

## D **Simulation** (p 94)

*'This was good – I brought in using the random number button on the calculator to aid simulations.'*

It is important for pupils to realise that each run of a simulation is like one trial within an experiment. Although running a simulation once may itself be a lengthy process, it has to be run many times to get enough data to estimate probabilities. Class results can be pooled. If individual pupils get very different results, then this is because their numbers of trials were too small.

**D1** The theoretical probability that 6 or more packets will be needed to collect the set is $\frac{31}{81}$.

**D2** The theoretical probability that 4 or more births are needed before a family has both girls and boys is $\frac{1}{4}$.

## ⅇ How often? (p 95)

◊ 'Expectation' is a technical term in probability theory. Here, however, it is used more informally.

### ⅈ Relative frequency (p 91)

**B1** $\frac{16}{50}$ or $\frac{8}{25}$ or 0.32

**B2** (a) $\frac{12}{40}$ or $\frac{3}{10}$ or 0.3 (b) $\frac{18}{40}$ or $\frac{9}{20}$ or 0.45
(c) $\frac{10}{40}$ or $\frac{1}{4}$ or 0.25

**B3** (a) $\frac{32}{50}$ or $\frac{16}{25}$ or 0.64 (b) $\frac{18}{50}$ or $\frac{9}{25}$ or 0.36
The coin does seem to be unfair.

**B4** (a) Fred counted 124 papers altogether.
*Mirror* 0.18   *Sun* 0.15
*Express* 0.10   *Mail* 0.14
*Star* 0.09   *Telegraph* 0.12
*Times* 0.09   *Guardian* 0.08
*Independent* 0.06
They add up to 1.01 because of rounding errors.
(b) 0.15

**B5** (a) $\frac{62}{250}$ or 0.248 (b) $\frac{44}{250}$ or 0.176

### ℂ Estimating probabilities (p 93)

**C1** The pupil's estimate. The theoretical probability is approximately 0.43.

**C2** (a) The pupil's guess
(b) The pupil's estimate. The theoretical probability is 0.375.
(c) The pupil's estimate. The theoretical probability is approximately 0.53.

### ⅇ How often? (p 95)

**E1** (a) About 50 times (b) About 150 times

**E2** (a) $\frac{40}{100}$ or $\frac{2}{5}$ or 0.4 (b) About 160 times

**E3** (a) $\frac{44}{100}$ or $\frac{11}{25}$ or 0.44 (b) About 110 times

**E4** About 350 times

### What progress have you made? (p 95)

**1** $\frac{35}{50} = \frac{7}{10} = 0.7$

**2** About 80 times

### Practice booklet

### Section B (p 44)

**1** 0.85 or $\frac{17}{20}$

**2** (a) $\frac{11}{30}$ (b) $\frac{10}{30}$ (c) $\frac{9}{30}$

**3** (a) $\frac{21}{40}$ (b) $\frac{19}{40}$
Naomi seems slightly better.

**4** Ready salted 0.13, Salt and vinegar 0.20
Cheese and onion 0.23, Prawn cocktail 0.25,
Beef 0.08, Pickled onion 0.11

### Section E (p 45)

**1** (a) About 60 times (b) About 30 times
(c) About 60 times

**2** (a) About 83 times (b) About 250 times
(c) About 167 times (d) About 167 times
(e) About 417 times

**3** (a) $\frac{23}{70}$ or approximately 0.33
(b) About 66 times

**4** (a) $\frac{83}{235}$ or 0.35 to 2 d.p.
(b) Ready salted 353,
Salt and vinegar 221,
Cheese and onion 174,
Prawn cocktail 115,
Beef 81,
Pickled onion 55

# ⑮ Bearings

**Essential**

Angle measurer/protractor
Sheets 202 to 205

**Practice booklet** page 46

## Ⓐ **Direction** (p 96)

> Sheet 202 (each pupil may need two copies)

◊ Before using the examples on the page, you could review compass
directions: N, E, NE, etc. Relate each of them to an angle measured
clockwise from N as zero, and emphasise that the angle corresponding to,
say, NE is written as 045°.

The bearings shown on the page are 125° for the direction of the tanker
and 250° for the direction of the ferry.

Note: Rathlin Island (sheet 202) is in Northern Ireland.

## Ⓑ **On the moors** (p 98)

> Sheet 203

The area used for this section is part of Exmoor National Park.

**B10** The bearing of a reverse direction is called a 'back bearing'. It is either
180° greater or 180° less than the bearing of the forward direction.

**\*B11** This problem can be solved using back bearings. However, in practice
navigators do not need to calculate a back bearing. They just use the
protractor 'backwards', putting the scale mark against the position of the
observed point (e.g. a lighthouse) and drawing a line back through the
centre mark.

*'If the bearing of L from me is 036°,
then I must be somewhere on this line.'*

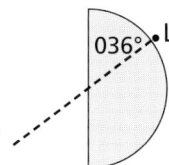

# Ⓒ **Bearings jigsaw puzzle** (p 99)

| Sheets 204 and 205 |
|---|

◊ This puzzle gives plenty of practice in bearings and some scope for logical thinking.

## Ⓐ **Direction** (p 96)

**A1** (a) 008°          (b) 032°

**A2 to A6**

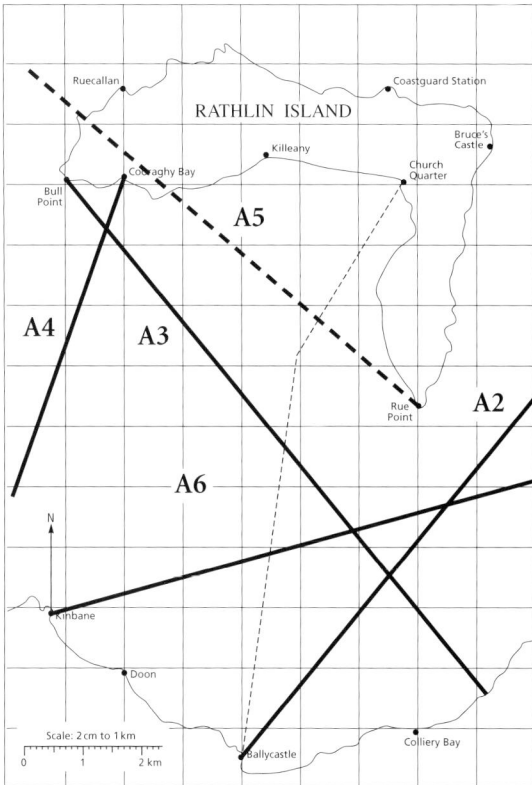

© Crown copyright

**A5** On her left

## A7 to A12

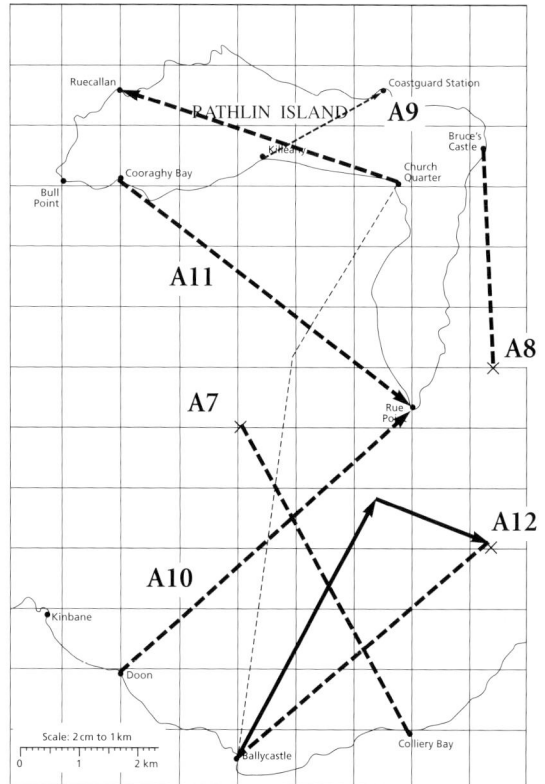

© Crown copyright

**A9** 062°

**A10** 049°

**A11** (a) 127°          (b) 288°

**A12** 5.7 km on a bearing of 050°

## A13 and A14

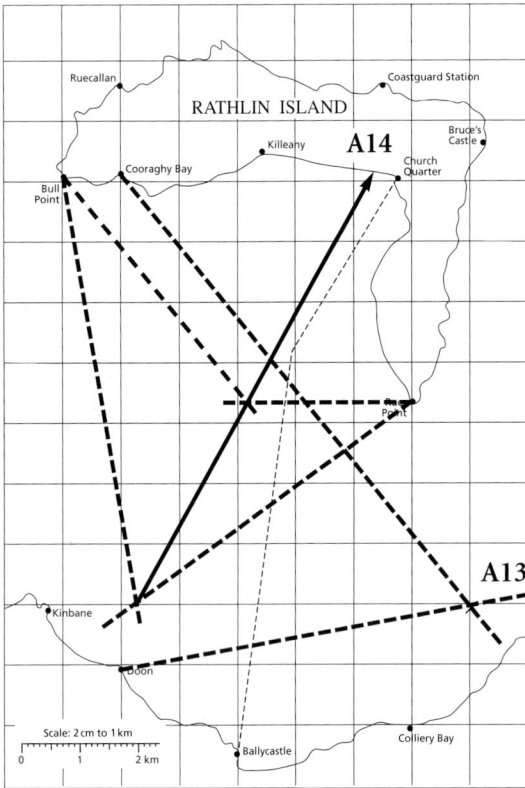

© Crown copyright

## B On the moors (p 97)

**B1** 12 cm

**B2** (a) 18 cm    (b) 11.2 cm    (c) 6.4 cm
     (d) 19.6 cm    (e) 3.2 cm

**B3** 4 km

**B4** (a) 2 km    (b) 3.75 km    (c) 1.8 km
     (d) 2.6 km    (e) 1.35 km

## B5 to B9

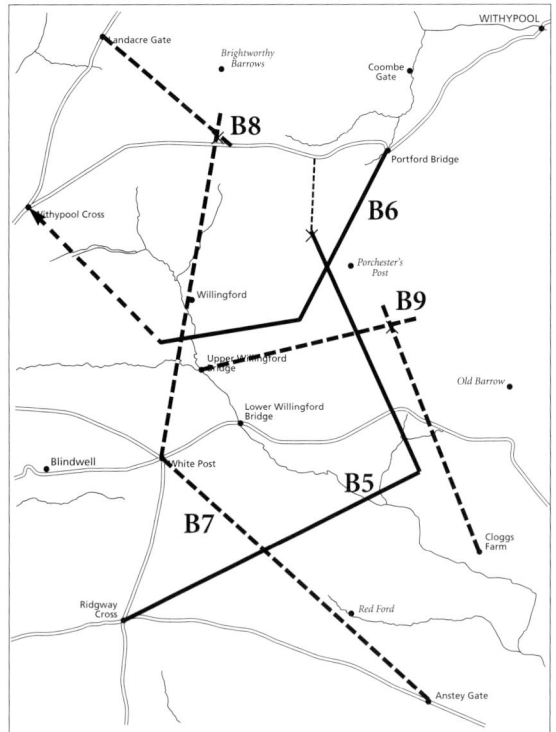

© Crown copyright

**B5** (b) 0.6 km

**B6** (b) 315°

**B7** (a) 312°        (b) 3 km

**B10** (a)

| RC to CF | **079°** | CF to RC | **259°** |
|---|---|---|---|
| CG to OB | **161°** | OB to CG | **341°** |
| PP to W | **258°** | W to PP | **078°** |
| WC to WP | **151°** | WP to WC | **331°** |
| RF to OB | **036°** | OB to RF | **216°** |

(b) The difference between them is 180°.

*B11

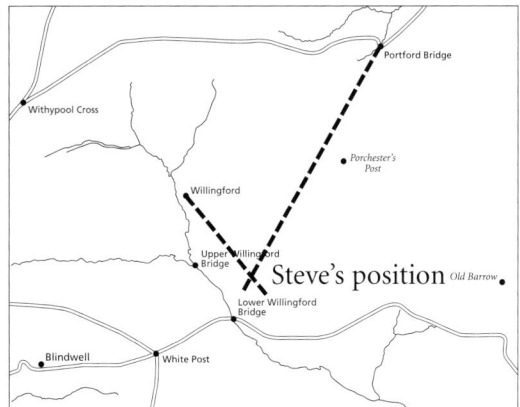

© Crown copyright

## ℂ Bearings jigsaw puzzle  (p 99)

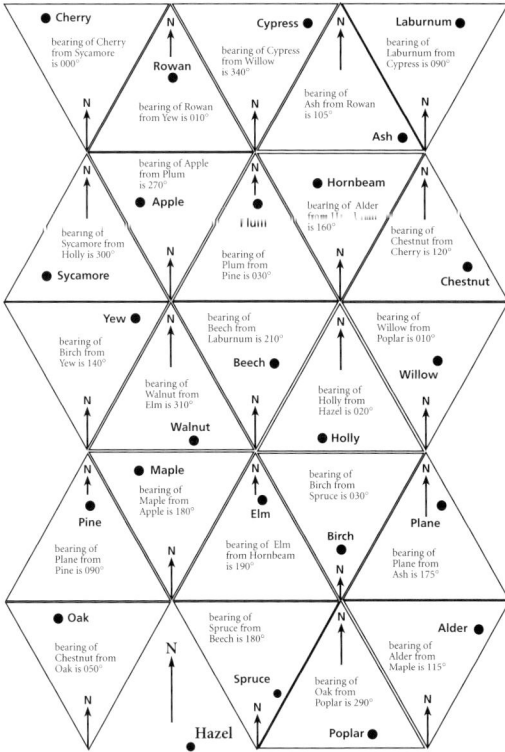

## What progress have you made?  (p 99)

**1** (a)  124°          (b)  281°

## 2 and 3

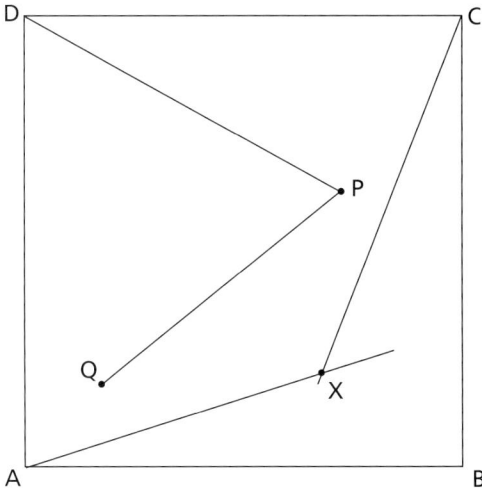

## Practice booklet

## Section A  (p 46)

**1** (a)  090°      (b)  058°      (c)  000°

(d)

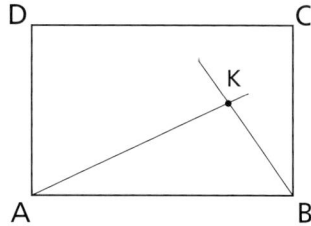

(e)  110°      (f)  25.6 km

**2** (a)  310°      (b)  190°

**3** (a)  015°      (b)  330°      (c)  105°

(d)  195°      (e)  240°

# ⑯ Forming equations

This unit introduces pupils to forming a mathematical expression from a situation presented in words. Pupils then go on to solve a resulting equation.

The emphasis is on using algebra, rather than solving problems. All the problems might be solved more or less easily without algebra. It is therefore important to 'come clean' with the pupils and tell them that the unit will help them gain and practise skills that will be invaluable later.

| | | |
|---|---|---|
| T | p 100 **A** Expressions | Forming a mathematical expression from a situation presented in words |
| T | p 103 **B** Equations | Forming and solving an equation with the unknown on one side |
| T | p 108 **C** Harder problems | Forming and solving an equation with the unknown on both sides |

---

**Practice booklet** pages 47 to 50

---

## Ⓐ **Expressions** (p 100)

◊ This section is about forming expressions in a single variable. You might approach this with the class by using the picture on page 100 as a stimulus. For example:

'The farmer has some bulls (but we can't see how many). He has twice as many cows. How could we write this using mathematical symbols?' leading to: 'Suppose the farmer has $n$ bulls. Then he has $2n$ cows.'

*'I used this picture and made up lots of statements to put into expressions. Worked well.'*

You might go on to ask for suggestions for 'translating' such sentences as:
'There are six fewer chickens than bulls.'
'There are twice as many ducks as chickens' and so on.

The unit has problems leading to expressions like
$2n$, $3n + 2$, $\frac{n}{2} + 2$, $3(n + 4)$ etc, so ingenuity will be called for in devising sentences to cover different expressions!

◊ One of the confusions that can arise when forming expressions is linguistic. It is important that pupils understand that when we say 'There are $n$ bulls and twice as many cows. How many cows are there?' the 'There are ...' and 'How many ...' in this question do not mean that we know the actual numbers. They are shorthand, meaning 'If $n$ stands for the number of bulls ... write an expression for the number of cows.' But we do use the shorthand, and it is important that pupils see it, and do not think that there is no answer to 'How many cows are there?'

◊ Another misconception is believing that we use a letter to stand for a word, rather than a number. So if there are three rabbits to every one fox, pupils will often write $3r = f$ (and, perhaps, '$r$ stands for rabbits and $f$ for foxes'). The point that the letters stand for the **number** of rabbits, etc., needs constant emphasis.

◊ If pupils find forming an expression difficult, suggest that they do the equivalent operations with numbers, think about what they have done, and then try to form the expression.

◊ There is much scope for discussion of equivalent expressions when pupils may come up with apparently quite different, but correct, answers to problems.

## B **Equations** (p 103)

◊ In this section, pupils form an expression and then use it to solve an equation. You can use the pizza problem to introduce the ideas to the whole class. A suggested problem might be:

> Nick bought a pizza and cut it into three.
> Piece B weighed 15 grams more than A.
> Piece C weighed 6 grams less than A.
> The whole pizza weighed 420 grams.
> How much did each piece weigh?

Leading to:

'I did this using the pizza picture – worked well.'

> Suppose A weighs $x$ grams.
> Then B weighs $x + 15$ grams, and C weighs $x - 6$ grams.
> So all three pieces weigh $x + x + 15 + x - 6 = 3x + 9$ grams
> So $\quad 3x + 9 = 420$
> $\qquad\quad 3x = 411$
> $\qquad\quad\, x = 137$
> Piece A weighs 137 g; B weighs 152 g; C weighs 131 g.

Check in the original problem: 137 g + 152 g + 131 g = 420 g.

You might want to go on to ask the class to discuss the solution of similar problems before they begin the questions in the section.

## C **Harder problems** (p 108)

◊ In this section, pupils form an equation where the unknown appears on both sides. When solving problems of this type, it is important to pick the right thing to be the unknown. You can use the ribbon problem as an introduction to the whole class to illustrate this.

> I make bows from two reels of ribbon, one yellow and one red. Both reels have the same length of ribbon on them at the start.
>
> I make 7 yellow bows and there is 100 cm left over on the reel. I make 10 red bows and there is 10 cm of red ribbon left over on the reel.
>
> How much ribbon was there on each reel originally?

If we choose what we are asked to find (the length of ribbon on a reel) as the unknown, the problem is much harder to solve.

So we choose the length of ribbon in one bow as the unknown.

> Suppose there are $x$ cm of ribbon in a bow.
> Then 7 yellow bows use $7x$ cm. Together with the 100 unused cm, there was originally $7x + 100$ cm on the yellow reel.
> Similarly there was originally $10x + 10$ cm on the red reel.
> So $7x + 100 = 10x + 10$; $90 = 3x$; $x = 30$.
> So there is $7 \times 30 + 100$ cm $= 310$ cm on a reel.
> Check: the red reel contains $10 \times 30 + 10$ cm $= 310$ cm.

Ⓐ **Expressions** (p 100)

Equivalent expressions to any answer are perfectly acceptable.
At this stage you do not need to worry the pupils about using correct units.

**A1** (a) $3n$  (b) $n + 5$  (c) $\frac{n}{2}$ or $n \div 2$

**A2** (a) $3n$  (b) $4n$
   (c) $n - 3$  (d) $2n + 3$

**A3** (There are several ways of expressing each of these.)
   (a) There are twice as many tigers as lions.
   (b) There are 8 more pumas than lions.
   (c) There are two less leopards than lions.
   (d) There are half as many jackals as lions.
   (e) There is one more hyena than there are jackals.

**A4** The pupil's sentences

**A5** (a) $4c$  (b) $4(c - 6)$

**A6** $9(m - 1) + 4$ or $9m - 5$

**A7** (a) $100x$  (b) $100x + 80$
   (c) $12x + 18$
   (d) $x - 10$ (trucks);
      $100(x - 10) + 80$ or $100x - 920$ (tonnes)

**A8** (a) $\frac{850}{k}$  (b) $\frac{800}{k - 1}$

**A9** (a) $4n$  (b) $2n - 2$ and $2n + 2$

**A10** $\frac{n - 20}{10}$

**A11** (a) $\frac{l}{4}$  (b) $\left(\frac{l}{4}\right)^2$

*A12 (a) (i) $2(a + 6)$  (ii) $6a$
   (b) (i) $\frac{p - 16}{2}$  (ii) $4(p - 16)$

**Equations** (p 103)

Pupils should show that they have checked each of their answers.

**B1** (a) £64$n$      (b) $64n = 352$

   (c) $n = 5.5$,
so he buys 5.5 tonnes of rock.

**B2** (a) £7.6$x$

   (b) $7.6x = 57$; $x = 7.5$,
so Jill buys 7.5 metres of material.

**B3** (a) $8d + 13$

   (b) (i) $8d + 13 = 141$

      (ii) $d = 16$; each box holds 16 discs.

**B4** (a) £$(s + 12)$    (b) £$(2s + 12)$

   (c) $2s + 12 = 73$; $s = 30.5$,
so Susan spends £30.50
and Jon spends £42.50.

**B5** (a) Alan: £$x$       Becky: £4$x$
Colin: £$(x + 4)$    Debbie: £2$x$
Eddie: £$(4x - 4)$   Fran: £3$x$

   (b) $x + 4x + x + 4 = 13$; $6x + 4 = 13$;
$x = 1.5$.
Alan gets £1.50;   Becky gets £6;
Colin gets £5.50;  Debbie gets £3;
Eddie gets £2;     Fran gets £4.50.

**B6** (a) Apple: $c$    Pear: $c + 5$
Peach: $5c$   Banana: $c + 10$
Satsuma: $\frac{c}{2}$  Apricot: $5c - 5$

   (b) $c + 5 + 5c - 5 = 60$; $c = 10$
An apple costs 10p, pear 15p,
peach 50p, banana 20p,
satsuma 5p, apricot 45p.

**B7** (a) $53n - 7$ (with reason)

   (b) $53n - 7 = 470$;
$n = 9$, i.e. they use 9 coaches.

**B8** (a) Caroline's age is $x - 4$
(she is **younger** than Robin).

   (b) $2x - 4 = 21$

   (c) $x = 12\frac{1}{2}$,
so Robin is $12\frac{1}{2}$ and Caroline is $8\frac{1}{2}$.

**B9** (a) The length is $2w$ cm and the perimeter is $6w$ cm.

   (b) $6w = 26$; $w = 4\frac{1}{3}$ or 4.33...

**B10** (a) The length is $5x$ cm and the area is $5xx$ or $5x^2$ cm$^2$

   (b) $5x^2 = 245$; $x^2 = 49$; $x = 7$

**B11** (a) £$(2c + 13.11)$

   (b) $2c + 13.11 = 43.21$; $c = 15.05$
A shirt costs £15.05.

**B12** (a) Suppose a pair of boots cost £$b$.
Trainers cost £$(b - 13.55)$.

   (b) $2b - 13.55 = 98.93$; $b = 56.24$
Boots cost £56.24.

**B13** (a) £$m + 29$

   (b) (i) £22$m$   (ii) £9$(m + 29)$

   (c) $22m + 9(m + 29) = 1594$; $m = 43$
A mini-disc Walkman costs £72,
a Walkman tape costs £43.

**B14** (a) $\frac{l}{2}$ cm

   (b) $4(\frac{l}{2} + l + l) = 370$
$l = 37$, so BC = 37 cm

**B15** (a) $y - 5$    (b) $y + 3$

   (c) $y + y - 5 + y + 3 = 40$; $y = 14$
Thomas is 14, John is 9, Susan is 17.

**B16** (a) $3n$     (b) £20$n$, £15$n$

   (c) £35$n$

   (d) $35n = 105$; $n = 3$
I have 3 £20 notes and 9 £5 notes.

\*B17 (a) $2f$  (b) $3f$  (c) $5f + 20f = 25f$

   (d) $3f = 33$; $f = 11$     (e) 275p

\*B18 Suppose the number is $n$.
Then $n + 12 + 2n - \frac{n}{2} = 37$
$2\frac{1}{2}n + 12 = 37$; $2\frac{1}{2}n = 25$; $n = 10$
I thought of 10.

\*B19 Suppose the number is $n$.
Then $n + \frac{n}{2} - \frac{1}{2}(n + \frac{n}{2}) = 75$
$\frac{1}{2}(n + \frac{n}{2}) = 75$; $(n + \frac{n}{2}) = 150$; $n = 100$
I thought of 100.

ℂ **Harder problems** (p 108)

Note: There are often correct alternative ways of working to the ones shown here.

**C1** (a) $3x + 27$      (b) $5x - 4$

(c) $3x + 27 = 5x - 4$; $x = 15\frac{1}{2}$
So Sarah and Juliet are both $15\frac{1}{2}$.

**C2** (a) Jim $3n + 18$; Harry $n + 32$

(b) $3n + 18 = n + 32$; $n = 7$

**C3** Zahir gets $8(x + 3)$;
Marisa gets $6(x + 13)$.
$8(x + 3) = 6(x + 13)$; $8x + 24 = 6x + 78$;
$x = 27$
They were both thinking of 27.

**C4** Suppose Alan's number is $a$.
Then Emdad's is $a + 2$. So:
$2a + 7 = 5(a + 2 - 1)$; $a = \frac{2}{3}$ (or 0.66...)
Alan thinks of $\frac{2}{3}$, Emdad thinks of $2\frac{2}{3}$.

**C5** (a) $x + 2$

(b) $4x - 5 = 2(x + 2) + 12$; $x = 10\frac{1}{2}$
So Ben is $12\frac{1}{2}$.

**C6** (a) We can compare any pair of sides.
So, for example,
$2x + 3 = 5x - 4$ gives $x = 2\frac{1}{3}$
(or 2.33…)
The length of the left-hand side is
then $2 \times 2\frac{1}{3} + 3 = 7\frac{2}{3}$ (or 7.66…)
The right-hand side is
$5 \times 2\frac{1}{3} - 4 = 7\frac{2}{3}$
and the base is $8 \times 2\frac{1}{3} - 11 = 7\frac{2}{3}$.

(b) If the triangle is equilateral then the length of the left side is equal to the length of the right side.
So $3x + 8 = 5x - 6$; $x = 7$
Left side = 29; right side = 29
The length of the base is then
$4 \times 7 + 2 = 30$.
So the triangle cannot be equilateral.

**C7** (a) Into the left of AB gas arrives at a rate of $25 + x$.
Out of the right of AB gas leaves at a rate of $7 + 4x$.

(b) These rates must be the same if AB is not gaining or losing gas.
So $25 + x = 7 + 4x$; $x = 6$

(c) 31 cubic metres per minute

**C8** As in question C7, we have
$52 + 3x = 17 + 2x + 5x$.
So $x = 8.75$, and the flow along CD is $52 + 3 \times 8.75 = 78.25$ (cubic metres per minute).

**\*C9** Suppose the shelves in the bookcase are each $s$ cm long. The 5 shelves use $5s$ cm, and there is 20 cm over. So the plank was $5s + 20$ cm long at the start.
Each shelf of the cabinet is $s + 10$ cm.
3 of these shelves use $3(s + 10)$ cm;
with the 50 cm over that plank was
$3(s + 10) + 50$ cm long.
So $5s + 20 = 3(s + 10) + 50$; $s = 30$.
So each plank was $5 \times 30 + 20 = 170$ cm long.

**What progress have you made?** (p 110)

**1** $3a + 4(a + 2)$ or $7a + 8$ pence

**2** Today I spend $t$ minutes.
Yesterday I spent $(t - 20)$ minutes.
Tomorrow I expect to spend $2t$ minutes.
So $t + t - 20 + 2t = 140$; $t = 40$
So yesterday I spent $40 - 20 = 20$ minutes on my homework.

**3** Suppose they both think of $n$.
William does $4n - 3$.
Brian does $2(n + 5)$.
So $4n - 3 = 2(n + 5)$; $n = 6\frac{1}{2}$

**Practice booklet**

**Section A** (p 47)

**1** (a) (i) Ben has $3x$ sweets.
(ii) Chloe has $(x - 3)$ sweets.
(iii) Daniel has $\frac{1}{2}x$ or $\frac{x}{2}$ sweets.

(b) 'Tina has one more sweet than Ben' or 'Tina has three times the number of sweets Alan has, plus one.'

2  $6n - 6$ are left.

3  (a)  $12s$

   (b)  $12(t - 1) + 4$ or $12t - 8$

## Section B (p 48)

1  (a)  William has $q - 6$ mice.

   (b)  $q + q - 6 = 84$

   (c)  Iain has 45 mice.
        William has 39 mice.

   (d)  The pupil's check

2  (a)  The perimeter of the rectangle is
        $(x + 5) + (x + 5) + x + x$ cm
        $= 4x + 10$ cm

   (b)  $4x + 10 = 74$

   (c)  $x = 16$
        The sides are 16 cm and 21 cm long.

3  (a)  Rehana has $\frac{n}{2}$ or $\frac{1}{2}n$ bulbs.

   (b)  Roisin has $n + 100$ bulbs.

   (c)  $n + \frac{n}{2} + n + 100 = 1000$,  $n = 360$

   (d)  Imran has 360 bulbs,
        Rehana has 180, Roisin has 460.

4  (a)  $3x + 30 = 180$

   (b)  $x = 50$

   (c)  The angles are 50°, 50° and 80°.

5  (a)  (i)   Six glasses cost $6g$ pence.

        (ii)  Six plates cost $6(g + 20) =$
              $6g + 120$ pence

        (iii) $6g + 6g + 120 = 12g + 120$

   (b)  $6g + 6g + 120 = 540$
        or $12g + 120 = 540$

   (c)  $g = 35$
        A glass costs 35p and a plate costs 55p.

6  $y + y + 3 + \frac{y}{2} = 38$

   $y = 14$

   Amy is 14 years, Zoe is 17 years old and Mike is 7 years old.

## Section C (p 50)

1  The sides are 7 cm and 54 cm.

2  (a)  Alice is $(k - 3)$ years old.

   (b)  Danny is 15 years old.

3  (a)  1.50 euros

   (b)  They both start with 14.50 euros.

4  They both thought of $4\frac{1}{2}$.

# Review 2 (p 111)

**1** (a) $x = 4$

  (b) 90° clockwise, centre $(0, 0)$

  (c) 180°, centre $(4, 5)$

  (d) $\begin{bmatrix} 10 \\ 8 \end{bmatrix}$

**2** (a) Reflection in $y = 5$

  (b) Reflection in $x = y$

  (c) Translation with vector $\begin{bmatrix} 10 \\ 0 \end{bmatrix}$

  (d) Translation with vector $\begin{bmatrix} -8 \\ -10 \end{bmatrix}$

  (e) Rotation 90° anticlockwise, centre $(0, 0)$

  (f) Rotation through 180°, centre $(-5, -4)$

  (g) Reflection in $y = -x$

  (h) Rotation 90° anticlockwise, centre $(5, 5)$

**3** (a) To increase a quantity by 16%, multiply it by **1.16**.

  (b) £566

**4** $a$   $y = x + 3$

  $b$   $y = x - 4$

  $c$   $y = 2x + 4$

  $d$   $y = 2x - 7$

  $e$   $y = \frac{1}{2}x + 5$

  $f$   $y = -x - 4$

  $g$   $y = -2x - 2$

**5** (a) (i) 0.11     (ii) 0.80     (iii) 0.09

  (b) There are four faces it could land on to be in position B.

**6** (a) $-7x + 11$ or $11 - 7x$

  (b) $-2x + 5$ or $5 - 2x$

  (c) $-3x + 4$ or $4 - 3x$

  (d) $6x - 3y - 1$

  (e) $-5x - 3y + 13$

  (f) $3x - 3y - 2$

**7**

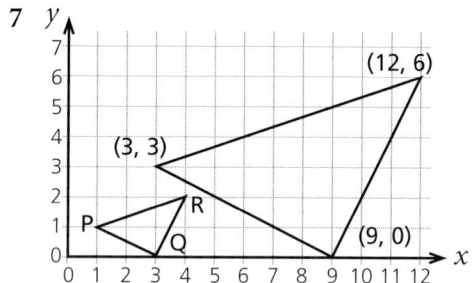

**8** (a) To reduce a quantity by 7%, multiply it by **0.93**.

  (b) £4020

**9** (a) $20 - n$

  (b) $5n + 3(20 - n) = 2n + 60$

  (c) $2n + 60 = 74$ leading to $n = 7$ so she has 7 white rabbits.

**10**

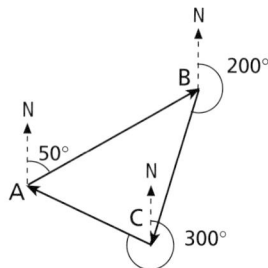

Angle ABC $= 30°$

Angle BCA $= 80°$

Angle CAB $= 70°$

**11** (a) 14%       (b) 11%

**12** (a) (i) $4n$   (ii) $4n - 8$   (iii) $n + 8$

  (b) $4n - 8 = 2(n + 8)$ leading to $n = 12$

**Mixed questions 2** (Practice booklet p 51)

1 (a) $y = 3x - 2$

(b) $y = \frac{1}{2}x + 1$

(c) $y = -\frac{1}{2}x + 4$

(d) $x = \frac{3}{2}x + 5$

2 (a) Reflection in the line $y = 2$

(b) Rotation $90°$ anticlockwise about $(0, 2)$

(c) Rotation $90°$ anticlockwise about $(0, 2)$

(d) Translation by the vector $\begin{bmatrix} -7 \\ 1 \end{bmatrix}$

(e) Rotation of $180°$ about $(1, 2)$

(f) Rotation of $180°$ about $(4.5, 1.5)$

3 2% increase

4 (a) 0.15    (b) £7.50

5 First section    15 minutes
Second section    38 minutes
Third section    76 minutes
Fourth section    66 minutes

6 (a) $220°$    (b) $284°$

7 (a) P goes with graph 1.
Q goes with graph 3.
R goes with graph 2.

(b)    **Graph for track S**

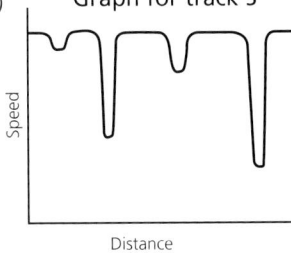

Speed
Distance

# ⑰ Ratio and proportion

---

**Practice booklet** pages 53 and 54

---

## 🄐 Some problems (p 114)

**T**

*'Very useful. I gave each problem to a group of four, who then explained their method to the class.'*

◊ These problems bring out ideas taken up later in the unit.

Pupils may suggest additive methods. For example, for the third problem pupils may suggest both ÷ 4, × 7 and treating 7 as 4 + 2 + 1. If the multiplying method does not come from the pupils then you can use the examples to introduce and emphasise it.

◊ The answers to the questions are as follows.

**1**   200 litres

**2**   150 g flour, 375 ml milk, 45 g margarine, 3 eggs

**3**   525 g damsons, 35 ml water, 175 g sugar, 35 g custard powder, 262.5 ml milk

**4**   100 g butter, 1250 g potatoes, 1875 ml vegetable stock, 250 g cheese, 7.5 (or 7 or 8) leeks

**5**   22.5 acres

**6**   12.5 litres red and 7.5 litres yellow

## 🄑 Direct proportion: multipliers (p 115)

## 🄒 Direct proportion: unitary method (p 116)

## $\mathbb{D}$ **Graphs and equations** (p 117)

◊ You may need to revise gradient and equations of straight lines.

## $\mathbb{E}$ **Conversion graphs** (p 118)

◊ Pupils should appreciate that apart from $(0, 0)$ only one point is needed to fix the graph. To achieve greater accuracy the second point should be as far from the origin as possible.

## $\mathbb{F}$ **Ratios and fractions** (p 120)

◊ The questions ring the changes on the different comparisons that can be made using either fractions or ratios. For example if an amount is shared unequally between two people, fractions or ratios can be used to compare each share with the other share or each share with the total.

◊ It is helpful to draw a diagram showing a ratio or fraction. For example when the ratio $3:2$ is shown like this $\boxed{\phantom{xx}3\phantom{xx}\mid\phantom{xx}2\phantom{xx}}$ it can be seen easily that the first part is $\frac{3}{5}$ of the whole, or $1\frac{1}{2}$ times the second part, etc.

**F8** This question links together the different ways of representing proportion that have been met in the unit.

---

### $\mathbb{B}$ **Direct proportion: multipliers** (p 115)

**B1** 7.5 litres

**B2** 80 seconds

**B3** 55 litres

**B4** 13.5 m

### $\mathbb{C}$ **Direct proportion: unitary method** (p 116)

**C1** 385 copies

**C2** 5280 words

**C3** 112 g

**C4** 192 kg

**C5** 17.9 kg

**C6** 57.7 m$^2$

**C7** (a) 3.6 litres   (b) 126 m$^2$

**C8** £6.06

### $\mathbb{D}$ **Graphs and equations** (p 117)

**D1** (a)

| $h$ | 0 | 3 | 5 | 8 | 10 | 12 |
|---|---|---|---|---|---|---|
| $s$ | 0 | 4.5 | **7.5** | **12** | **15** | **18** |

(b) The pupil's graph

(c) $s = 1.5h$

**D2** (a) $n = 3.5t$   (b) $n = 119$

**D3** The points do not lie on a straight line through $(0, 0)$.

**D4** (a) The pupil's graph

(b) $M = 0.8V$

(c) (i) $M = 57.6$   (ii) $V = 87.5$

## E Conversion graphs (p 118)

**E1** (a) About 21 km  (b) About 19 miles
(c) About 38 km  (d) About 14 miles

**E2** (a) One way is to convert 24 miles and double the result.
(b) About 62 or 63 miles

**E3** (a) The pupil's graph through (25, 46)
(b) (i) About 9 km per hour
(ii) About 36 km per hour
(iii) About 11 knots
(iv) About 19 knots

**E4** (a) About 96 km  (b) About 101 km
(c) About 63 miles

**E5** (a) The pupil's graph through (50, 130)
(b) (i) About 96 km$^2$
(ii) About 31 square miles

**E6** (a) The pupil's graph through (125, 190)
(b) (i) About $74.50
(ii) About £22
(iii) About $105
(c) $1.52

## F Ratios and fractions (p 120)

**F1** (a) $\frac{1}{2}$  (b) $\frac{2}{3}$  (c) $\frac{1}{3}$  (d) 2

**F2** (a) $\frac{2}{5}$  (b) $\frac{2}{3}$  (c) $\frac{3}{5}$  (d) $1\frac{1}{2}$

**F3** (a) 1:3  (b) 5

**F4** 24

**F5** $\frac{5}{9}$

**F6** (a) 3:1  (b) 2:1  (c) 2:3

**F7** 3:7

**F8** There are 4 different shades.
blue : yellow = 2:3  A, E, K, O
blue : yellow = 3:2  G, H, J, L, M
blue : yellow = 3:5  B, C, F
blue : yellow = 5:3  D, I, N

### What progress have you made? (p 122)

**1** 75 minutes

**2** 36 kg

**3** (a) 41.7p  (b) £6.58

**4** $D = 1.25T$

**5** 5:3

## Practice booklet

## Sections B and C (p 53)

**1** 2 hours

**2** 50 minutes

**3** (a) 16 m$^2$  (b) 72 m$^2$

**4** £28.70

**5** £14.39

## Sections D and E (p 54)

**1** $C = 2.5W$

**2** The pupil's graph through (60, 50)
(a) 18 square metres
(b) 17 square yards

## Section F (p 54)

**1** (a) (i) $\frac{4}{5}$  (ii) $\frac{4}{9}$
(b) 20

**2** 11:5

# ⑱ No chance!

---

**Essential**

Tetrahedral (four sided) dice (one per pair)
or you can use ordinary dice and ignore throws of 5 or 6.
Counters (two colours)
Sheet 206

**Practice booklet** pages 55 and 56

---

## Ⓐ **Beat the teacher** (p 123)

Thinking about these games leads to a need for systematically listing the equally likely outcomes. You may decide to move straight to using a grid; however, some of questions in section B involve three throws, where a grid is not so useful. Grids are introduced in section C.

◊ You could play one or more of these games with the class, but the results would be inconclusive. In each case all the equally likely outcomes need to be listed.

**Coin game**

The teacher is being unfair. Of the four possible outcomes HH, HT, TH, TT, the teacher wins on two and the pupil on one; the fourth outcome leads to another round where the teacher again has the advantage.

**Beat that!**

This game is fair. Omitting the 6 from the teacher's throw, there are 30 equally likely outcomes of the pair of throws. Half of these lead to a win for the pupil.

**Dice difference**

This game is unfair. Of the 36 equally likely outcomes, 16 give a difference of 0 or 1.

## B  All the outcomes (p 124)

**\*B5**  In reality the probability of a male birth is about 0.515.

## C  Easier ways of listing (p 125)

> Tetrahedral dice (one per pair), or you can use ordinary dice
> and ignore throws of 5 or 6
> Counters (two per pair, different colours)
> Sheet 206 (or pupils can easily draw the board)

### Pursuit

◊  Pupils should play the game a few times in order to see what decisions
need to be made as they play.

◊  Players need to know the relative likelihood of the different total scores
with the two four-sided dice. Introduce a grid for showing all the possible
outcomes and their totals.

Pupils should now be able to decide on the best move in the games
shown.

**C3**  As an extension, pupils could investigate the most likely total when
playing with two $n$-sided dice.

## D  Which is better? (p 127)

## E  How large, how many …? (p 128)

The methods described here show that probability has more uses than
simply playing games.

This is how one teacher did the experiments as class activities.

### Monte Carlo method

'I drew a 12 by 12 grid on the OHP and then drew a shape similar to the
one in the book. Each student generated a random coordinate using
12-sided dice. As each student read out their coordinate, I marked it on
the grid. We then discussed how we could use this information to
estimate the area. (The area was then estimated in the usual way – the
estimate obtained by sampling was remarkably accurate.)'

### Capture-recapture method

'I had a "population" of white cubes in a bag. I told the class that I had
previously "captured" 40 cubes and replaced them with black cubes. We
then selected a sample of 50 cubes (with replacement) and noted how

many were black. The challenge was to use this information to estimate how many cubes were in the bag. (We counted them afterwards to see how close the estimate was.)'

## B All the outcomes (p 124)

**B1** (a) The pupil's list of outcomes

(b) (i) $\frac{1}{9}$ (ii) $\frac{1}{3}$ or $\frac{3}{9}$ (iii) $\frac{1}{3}$ or $\frac{3}{9}$

**B2** (a) The pupil's list of 8 outcomes

(b) (i) $\frac{1}{8}$ (ii) $\frac{3}{8}$

(iii) $\frac{1}{4}$ or $\frac{2}{8}$ (iv) $\frac{1}{2}$ or $\frac{4}{8}$ (v) 0

**B3** (a) The pupil's list of 8 outcomes

(b) 8 (c) 4

**B4** (a) The pupil's list of 27 outcomes

(b) $\frac{1}{27}$ (c) $\frac{1}{9}$ or $\frac{3}{27}$

(d) $\frac{2}{3}$ or $\frac{18}{27}$ (e) $\frac{2}{9}$ or $\frac{6}{27}$

**\*B5** (a) (i) $\frac{1}{4}$ (ii) $\frac{1}{2}$

(b) The pupil's list of 16 possibilities

(c) (i) $\frac{1}{16}$ (ii) $\frac{5}{16}$ (iii) $\frac{6}{16}$ or $\frac{3}{8}$

**\*B6** (a) (i) 16 (ii) 32

(b) $\frac{1}{128}$

(c) For $n$ spinners there are $3^n$ outcomes.

## C Easier ways of listing (p 125)

**C1** (a) 0 (b) $\frac{1}{8}$ or $\frac{2}{16}$

(c) $\frac{13}{16}$ (d) $\frac{1}{16}$

**C2** (a) $\frac{1}{6}$ or $\frac{6}{36}$ (b) $\frac{1}{12}$ or $\frac{3}{36}$ (c) $\frac{1}{36}$

(d) $\frac{1}{2}$ or $\frac{18}{36}$ (e) $\frac{15}{36}$ (f) $\frac{1}{12}$ or $\frac{3}{36}$

**C3** The pupil's grid

(a) 7 (b) 2 and 12

(c)

| Score | 2 | 3 | 4 | 5 | 6 | 7 | 8 | 9 | 10 | 11 | 12 |
|-------|---|---|---|---|---|---|---|---|----|----|----|
| Prob | $\frac{1}{36}$ | $\frac{2}{36}$ | $\frac{3}{36}$ | $\frac{4}{36}$ | $\frac{5}{36}$ | $\frac{6}{36}$ | $\frac{5}{36}$ | $\frac{4}{36}$ | $\frac{3}{36}$ | $\frac{2}{36}$ | $\frac{1}{36}$ |

**C4** (a) The pupil's grid

(b) (i) $\frac{2}{9}$ or $\frac{8}{36}$ (ii) $\frac{5}{18}$ or $\frac{10}{36}$

(iii) $\frac{1}{6}$ or $\frac{6}{36}$ (iv) $\frac{1}{6}$ or $\frac{6}{36}$

(v) 0

**C5** The pupil's grid

(a) $\frac{1}{5}$ or $\frac{5}{25}$ (b) $\frac{4}{25}$ (c) $\frac{9}{25}$

(d) $\frac{12}{25}$ (e) $\frac{1}{25}$ (f) $\frac{3}{5}$ or $\frac{15}{25}$

## D Which is better? (p 127)

**D1** The probabilities are

A 0.267 (3 d.p.) B 0.292 (3 d.p.) C 0.3

So C has best chance of winning.

**D2** The probabilities are

A 0.333 (3 d.p.) B 0.286 (3 d.p.)

So A is more likely.

**D3** The probabilities are

A 0.3125 (3 d.p.) B 0.255 (3 d.p.)

So A is more likely.

**D4** The probabilities are

A 0.722 (3 d.p.) B 0.618 (3 d.p.)

So A is more likely.

## E How large, how many, …? (p 128)

**E1** (a) $\frac{8}{40} = \frac{1}{5}$ (b) 160

### What progress have you made? (p 130)

1 (a) The pupil's list of 27 outcomes

   (b) (i) $\frac{1}{27}$     (ii) $\frac{1}{9}$ or $\frac{3}{27}$

      (iii) $\frac{2}{9}$ or $\frac{6}{27}$    (iv) $\frac{7}{9}$ or $\frac{21}{27}$

2 (a) The pupil's grid

   (b) (i) $\frac{1}{5}$ or $\frac{5}{25}$     (ii) $\frac{4}{25}$

3 The probabilities are

   A 0.625   B 0.722

   So B is more likely.

## Practice booklet

## Section B (p 55)

1 (a) The pupil's list of 16 outcomes

   (b) (i) $\frac{1}{4}$     (ii) $\frac{3}{4}$     (iii) $\frac{9}{16}$

2 (a)   14   16   46
        41   61   64
        19   49   96
        91   94   69

   (b) 12 different numbers can be made.

   (c) (i) $\frac{6}{12} = \frac{1}{2}$   (ii) $\frac{3}{12} = \frac{1}{4}$   (iii) $\frac{3}{12} = \frac{1}{4}$
      (iv) $\frac{3}{12} = \frac{1}{4}$   (v) $\frac{2}{12} = \frac{1}{6}$   (vi) $\frac{9}{12} = \frac{3}{4}$

## Section C (p 56)

1

| | | Pentagon spinner | | | | |
|---|---|---|---|---|---|---|
| | | 1 | 2 | 3 | 4 | 5 |
| | 1 | 2 | 3 | 4 | 5 | 6 |
| Square | 2 | 3 | 4 | 5 | 6 | 7 |
| spinner | 3 | 4 | 5 | 6 | 7 | 8 |
| | 4 | 5 | 6 | 7 | 8 | 9 |

2 (a) $\frac{3}{20}$         (b) $\frac{1}{20}$

   (c) $\frac{3}{20}$         (d) $\frac{4}{20} = \frac{1}{5}$

3 6 and 5 are the most likely scores.

4 (a)

| | | Hexagon spinner | | | | | |
|---|---|---|---|---|---|---|---|
| | | 1 | 2 | 3 | 4 | 5 | 6 |
| Triangle | 1 | 1 | 2 | 3 | 4 | 5 | 6 |
| spinner | 2 | 2 | 4 | 6 | 8 | 10 | 12 |
| | 3 | 3 | 6 | 9 | 12 | 15 | 18 |

   (b) 6 is the most likely score.

# ⑲ Strips

This unit develops simplifying linear expressions and follows on from 'Simplifying expressions'.
It includes expressions such as $7t + 3s - 2(4t - 5s)$.

| | |
|---|---|
| p 131 **A** Review | Revision of simplifying simple linear expressions, for example $6 - 8x + 3x - 1 + 2x$, $7 + 4x - 5 + 5y - 8y$ |
| p 132 **B** Addition strips | Consolidating work on simplifying and forming and solving equations |
| p 133 **C** Signs of change? | Simplifying linear expressions such as $2x + (7x - 5)$, $7x - (5 + 2x)$, $7x - (5y - 2x)$ |
| p 134 **D** Subtraction strips | Consolidating techniques from section C |
| p 135 **E** More simplifying | Simplifying linear expressions such as $2x + 3(x - 5)$, $7x - 2(5 + 3x)$, $7x - 3(5y - x)$ |

**Practice booklet** pages 57 to 59

## Ⓐ **Review** (p 131)

## Ⓑ **Addition strips** (p 132)

The context here is 'addition strips' where the number in each square is found by adding the numbers in the two previous squares.
The section ends with B7 which describes a calculating trick. It may well increase the dramatic effect if you demonstrate the trick to the whole class yourself.

◊ Question B2 shows how algebra is used to solve the addition strip problems in B3. In B3 although part (a) can be solved quite easily by trial and improvement the same is not true for the other strips.

You may want to point out one could choose a letter for the number in a different square, for example

| 3 | $n - 3$ | $n$ | $2n - 3$ | $3n - 3$ | $5n - 6$ (19) |
|---|---|---|---|---|---|

This leads to $5n - 6 = 19$ which gives $n = 5$.

**B4** Pupils may choose a letter for the number in the first square in these strips and work from there. However, the problems are most easily solved if the square to the right of the first number is chosen. Once that is known, the numbers at the beginning of the strip can be calculated by subtraction. Pupils could solve the problems by labelling different squares and compare their methods.

**B7** This trick works best if you can multiply by 4 quickly (by doubling and doubling again) to give the total.

As an extension, ask pupils to devise a similar trick to use with a ten square strip (in this case you need to multiply the 7th number by 11 to find the total).

## ℂ **Signs of change?** (p 133)

◊ One approach is described below.

Ask pupils to calculate the following (shown on page 133) and to compare their results with others.

$$10 + (5 + 1) \quad 10 + (5 - 1) \quad 10 - (5 + 1) \quad 10 - (5 - 1)$$

As a whole class, discuss the results and methods used.

Pupils could now consider the results of the calculations without the brackets, working from left to right.

$$10 + 5 + 1 \quad 10 + 5 - 1 \quad 10 - 5 + 1 \quad 10 - 5 - 1$$

They should discover that, when adding an expression in brackets, removing the brackets makes no difference but this is not the case when subtracting an expression in brackets. In groups, pupils could try to devise their own explanations for this.

$10 - (5 + 1) = 10 - 5 - 1 = 4$ is usually easier for pupils to grasp – they can often understand that taking away two numbers added together is equivalent to taking away the first number followed by taking away the second.

$10 - (5 - 1) = 10 - 5 + 1 = 6$ is more difficult – it may help to think of $10 - (5 - 1)$ as taking away 1 less than 5, so if you start by taking away 5 (1 more than you need to) you need to then add 1 to make the calculation correct.

◊ Now consider algebraic expressions. Pupils could remove the brackets and simplify a set of expressions such as

$$7x + (6 - x) \quad 7x + (6 + x) \quad 7x - (6 + x) \quad 7x - (6 - x)$$

Ask them to calculate the value of each expression (the original and their simplified version) when $x = 2$ for example. Although this will not prove equivalence, it does suggest it.

Examples A, B and C before question C1 are reminders for pupils.

◊ It is worth relating the approach above to the rules for multiplying directed numbers, replacing subtractions by additions of the corresponding negative number. For example:

$$10x - (5 - 2x) = 10x + {}^{-}1(5 + {}^{-}2x)$$
$$= 10x + {}^{-}5 + 2x$$
$$= 12x - 5$$

## Ⓓ Subtraction strips (p 134)

◊ You may wish to revise subtraction of negative numbers before pupils start this section.

**D3** Some pupils may notice that a subtraction strip is just an addition strip in reverse and so the following method can be used to find the missing numbers.

Choose a letter for the last blank square and fill in the strip.

| 2a + 24 (100) | a + 24 | a | 24 |
|---|---|---|---|

Since the number in the first square is 100 we know that $2a + 24 = 100$ which can be solved to give $a = 38$ .

## Ⓔ More simplifying (p 135)

◊ One way to approach simplifying an expression such as $15y - 4(5 + y)$ is to deal first with the multiplication and then remove the brackets.

$$15y - 4(5 + y) = 15y - (20 + 4y) \text{ etc.}$$

Examples A, B and C illustrate this method and pupils can discuss them.

◊ Alternatively, simplifying $15y - 4(5 + y)$ can be tackled in the following way.

$$15y - 4(5 + y) = 15y + {}^{-}4(5 + y) = 15y + {}^{-}20 + {}^{-}4y \text{ etc.}$$

## Ⓐ Review (p 131)

**A1** A and D (both equivalent to $p + 5$)
B and H (both equivalent to $9 - p$)
C and G (both equivalent to $13p + 5$)
E and F (both equivalent to $9 - 5p$)

**A2** (a) $4v + 1$    (b) $3 + 3w$    (c) $3x$
(d) $11$    (e) $3 - 4k$    (f) $5j - 5$
(g) $4 - h$    (h) $5 + 3g$    (i) $3 - 6f$
(j) $8 - 7e$    (k) $7d - 3$    (l) $5 - 7c$
(m) $3 - 4b$    (n) $2a$

**A3** (a) $3z + 6y$    (b) $7x + 3w$
(c) $7 + 6v + 2u$    (d) $6t - s$
(e) $3 + 9q - 8r$    (f) $n + 3p$
(g) $5l + 3 - 7m$    (h) $11 - 7k - 6j$
(i) $3h - 2g$    (j) $4e - 6f - 5$
(k) $7c - 1 - 5d$    (l) $9a - 13b$

## Ⓑ Addition strips (p 132)

**B1** (a)

| -1 | 8 | 7 | 15 | 22 |
|---|---|---|---|---|

(b)

| 5 | 7 | 12 | 19 |
|---|---|---|---|

(c)

| 9 | $n$ | $n+9$ | $2n+9$ | $3n+18$ |
|---|---|---|---|---|

(d)

| -3 | $k$ | $k-3$ | $2k-3$ |
|---|---|---|---|

(e)

| $4-p$ | $p$ | $4$ | $p+4$ |
|---|---|---|---|

(f)

| 6 | $m-6$ | $m$ | $2m-6$ | $3m-6$ |
|---|---|---|---|---|

**B2** (a)

| 3 | $p$ | $p+3$ | $2p+3$ | $3p+6$ |
|---|---|---|---|---|

(b) $p = 5$ gives 21 in the last square.

(c)

| 3 | 9 | 12 | 21 | 33 |
|---|---|---|---|---|

**B3** (a)

| 3 | 2 | 5 | 7 | 12 | 19 |
|---|---|---|---|---|---|

(b)

| 1 | 0.5 | 1.5 | 2 | 3.5 | 5.5 | 9 |
|---|---|---|---|---|---|---|

(c)

| 10 | -3 | 7 | 4 | 11 | 15 | 26 |
|---|---|---|---|---|---|---|

(d)

| 2 | -1.2 | 0.8 | -0.4 | 0.4 | 0 |
|---|---|---|---|---|---|

**B4** (a)

| -3 | 6 | 3 | 9 | 12 | 21 |
|---|---|---|---|---|---|

(b)

| 8.5 | 4.5 | 13 | 17.5 | 30.5 | 48 |
|---|---|---|---|---|---|

**B5**

| 6 | -5 | 1 | -4 | -3 | -7 | -10 |
|---|---|---|---|---|---|---|

**B6**

| 2 | 1.5 | 3.5 | 5 | 8.5 | 13.5 |
|---|---|---|---|---|---|

**B7** One way to explain this trick is to call the numbers in the first two squares $a$ and $b$ and complete the strip.

| $a$ | $b$ | $a+b$ | $a+2b$ | $2a+3b$ | $3a+5b$ |
|---|---|---|---|---|---|

Adding the expressions gives $8a + 12b$.
Multiplying the fifth expression by 4 gives $4(2a + 3b) = 8a + 12b$.

## Ⓒ Signs of change? (p 133)

**C1** A and H (both equivalent to $3a - 6$)
B and D (both equivalent to $a - 6$)
C and G (both equivalent to $3a - 8$)
E and F (both equivalent to $7a - 6$)

**C2** (a) $8x - 3$    (b) $3y - 4$    (c) $8 - z$
(d) $x + 6$    (e) $2y - 9$    (f) $z - 7$
(g) $8 - 3x$    (h) $6y - 6$    (i) $2z - 3$
(j) $2x - 2$    (k) $2 - 11y$    (l) $2z - 1$
(m) $8 - 2x$    (n) $3 - 5y$    (o) $-3t - 5$

**C3** (a) $4a + 2b$    (b) $7c - d$
(c) $8e - 4f$    (d) $4g + 3h$
(e) $5j - 6k$    (f) $5m - 5n$

(g)  $1 + 4q - 7p$    (h)  $3r - 3s$
(i)  $2t - 3u$    (j)  $4 + 5v - w$
(k)  $4x - 3y$    (l)  $3z - y$

## D  Subtraction strips (p 134)

**D1** (a)

| 100 | 61 | 39 | 22 | 17 | 5 | 12 |
|---|---|---|---|---|---|---|

(b)

| 12 | 5 | 7 | -2 | 9 | -11 |
|---|---|---|---|---|---|

(c)

| 3 | $a$ | $3-a$ | $2a-3$ | $6-3a$ | $5a-9$ | $15-8a$ |
|---|---|---|---|---|---|---|

(d)

| $b$ | 5 | $b-5$ | $10-b$ | $2b-15$ |
|---|---|---|---|---|

(e)

| 6 | $c$ | $6-c$ | $2c-6$ | $12-3c$ | $5c-18$ | $30-8c$ |
|---|---|---|---|---|---|---|

**D2** (a)

| 20 | $n$ | $20-n$ | $2n-20$ | $40-3n$ | $5n-60$ |
|---|---|---|---|---|---|

(b)  15    (c)  $n = 13$    (d)  $n = 12$

**D3** (a)

| 100 | 62 | 38 | 24 |
|---|---|---|---|

(b)

| 30 | 19 | 11 | 8 | 3 | 5 |
|---|---|---|---|---|---|

(c)

| 26 | 16 | 10 | 6 | 4 | 2 |
|---|---|---|---|---|---|

(d)

| 20 | 6 | 14 | -8 | 22 |
|---|---|---|---|---|

(e)

| 10 | 1 | 9 | -8 | 17 | -25 | 42 |
|---|---|---|---|---|---|---|

(f)

| 1 | 1.2 | -0.2 | 1.4 | -1.6 | 3 |
|---|---|---|---|---|---|

**D4**

| 1 | 7 | -6 | 13 | -19 | 32 | -51 | 83 |
|---|---|---|---|---|---|---|---|

**D5** One way to explain this trick is to call the numbers in the first two squares $a$ and $b$ and complete the strip.

| $a$ | $b$ | $a-b$ | $2b-a$ | $2a-3b$ | $5b-3a$ |
|---|---|---|---|---|---|

Adding the expressions gives $4b$. Multiplying the second expression by 4 gives $4b$.

**D6** yellow: $p - r$    green: $p - q$
purple: $q + r - p$

## E  More simplifying (p 135)

**E1**  A and B (both equivalent to $4p + 10$)
C and G (both equivalent to $8p - 10$)
D and F (both equivalent to $8p + 10$)
E and H (both equivalent to $4p - 6$)

**E2**  (a)  $9x - 1$    (b)  $8y + 7$    (c)  $9z + 14$
(d)  $3w - 15$    (e)  $10p - 5$    (f)  $7 - 7q$

**E3**  (a)  $11a + 15$  (b)  $2b - 2$    (c)  $11c + 1$
(d)  $2 - 10d$    (e)  $21e - 2$    (f)  6

**E4**  (a)  $14a + b$        (b)  $5c - 3d - 1$
(c)  $10e + 5f$        (d)  $3h$
(e)  $20k - 3j - 1$    (f)  $m - 11n$

### What progress have you made? (p 136)

**1**  (a)  $5 + 6q$        (b)  $5 - 2r$
(c)  $4s - t$        (d)  $2w - 2x + 5$

**2**  (a)  $3 + 5w$    (b)  $4x - 5$    (c)  $19 + 7y$
(d)  $8 + 14z$    (e)  $4c - 3d$

**3**  (a)

| 9 | $h$ | $9+h$ | $9+2h$ | $18+3h$ | $27+5h$ |
|---|---|---|---|---|---|

(b)  $h = 1$

**4**  (a)

| 5 | 1.5 | 6.5 | 8 | 14.5 | 22.5 | 37 |
|---|---|---|---|---|---|---|

(b)

| -4 | 6 | 2 | 8 | 10 | 18 |
|---|---|---|---|---|---|

**5**  (a)

| 15 | 12 | 3 | 9 | -6 | 15 |
|---|---|---|---|---|---|

(b)

| 5 | 3.2 | 1.8 | 1.4 | 0.4 | 1 |
|---|---|---|---|---|---|

**6**  (a)  $4m + 10$        (b)  $3b - 7c$

## Practice booklet

### Section A (p 57)

1 (a) $3k$     (b) $8 + 3m$     (c) $8 - 5n$
  (d) $10p - 3$    (e) $4 - 4q$     (f) $8$
  (g) $3s$         (h) $4t + 4$     (i) $8u - 5$
  (j) $4v - 4$

2 (a) $15u + 3v$       (b) $5y - 2x$
  (c) $8w - 4x - 7$    (d) $8a - 4b - 1$
  (e) $5c - 3d + 6$     (f) $6 - e - 2f$
  (g) $7 - 2g - 5h$     (h) $6m - 2k - 5$
  (i) $4 - 3p$         (j) $11 - 2p$

### Section B (p 57)

1 (a)

| 5 | 8 | 13 | 21 | 34 |
|---|---|----|----|----|

(b)

| ⁻2 | 7 | 5 | 12 | 17 | 29 |
|----|---|---|----|----|----|

(c)

| 6 | 3 | 9 | 12 | 21 |
|---|---|---|----|----|

(d)

| 1 | 7 | 8 | 15 | 23 |
|---|---|---|----|----|

2 (a)

| $p$ | 4 | $p+4$ | $p+8$ | $2p+12$ | $3p+20$ |
|-----|---|-------|-------|---------|---------|

(b)

| ⁻5 | $q$ | $q-5$ | $2q-5$ | $3q-10$ |
|----|-----|-------|--------|---------|

(c)

| 3 | $r-3$ | $r$ | $2r-3$ | $3r-3$ | $5r-6$ |
|---|-------|-----|--------|--------|--------|

(d)

| 9 | $s-9$ | $s$ | $2s-9$ |
|---|-------|-----|--------|

3 (a)

| $13-q$ | $q$ | 13 | $q+13$ | $q+26$ |
|--------|-----|----|--------|--------|

(b)

| $s-5$ | 5 | $s$ | $s+5$ |
|-------|---|-----|-------|

4 (a) $3a + 10$
  (b) 16            (c) $a = 9$

---

5 (a)

| 4 | 8 | 12 | 20 | 32 | 52 |
|---|---|----|----|----|----|

(b)

| 3 | 4.4 | 7.4 | 11.8 | 19.2 | 31 |
|---|-----|-----|------|------|----|

(c)

| 6 | ⁻4 | 2 | ⁻2 | 0 |
|---|----|---|----|---|

(d)

| 3.2 | 8.8 | 12 | 20.8 |
|-----|-----|----|------|

6 (a)

| ⁻3 | 5 | 2 | 7 | 9 | 16 |
|----|---|---|---|---|----|

(b)

| 4.3 | 5.2 | 9.5 | 14.7 | 24.2 | 38.9 | 63.1 | 102 |
|-----|-----|-----|------|------|------|------|-----|

### Section C (p 58)

1 A and C $(3x + 3)$     B and G $(x - 1)$
  D and F $(3x - 3)$     E and H $(x - 5)$

2 (a) $5x + 3$     (b) $3x - 2$     (c) $7x + 4$
  (d) $8 - x$      (e) $3x + 3$     (f) $3 - x$
  (g) $6x + 2$     (h) $x - 1$      (i) $3 - 6x$
  (j) $4 - 3x$     (k) $1 - x$      (l) $3x - 2$

### Section D (p 58)

1 (a)

| 21 | 12 | 9 | 3 | 6 | ⁻3 |
|----|----|---|---|---|----|

(b)

| 24 | 19 | 5 | 14 | ⁻9 |
|----|----|---|----|----|

(c)

| 5 | 8 | ⁻3 | 11 | ⁻14 |
|---|---|----|----|-----|

(d)

| 8 | 6 | 2 | 4 | ⁻2 | 6 |
|---|---|---|---|----|---|

**2** (a)

| 4 | a | 4−a | 2a−4 |
|---|---|---|---|

(b)

| b | 6 | b−6 | 12−b | 2b−18 |
|---|---|---|---|---|

(c)

| c+5 | 5 | c | 5−c |
|---|---|---|---|

(d)

| d+4 | d | 4 | d−4 |
|---|---|---|---|

(e)

| e | e−2 | 2 | e−4 | 6−e | 2e−10 |
|---|---|---|---|---|---|

**3** (a) $2x - 9$    (b) $1$     (c) $29.5$

**4** (a) $y = 13$    (b) $y = 10$    (c) $y = 10$

**5** (a)

| 58 | 31 | 27 | 4 |
|---|---|---|---|

(b)

| 42 | 29 | 13 | 16 | ⁻3 | 19 |
|---|---|---|---|---|---|

(c)

| 13 | 7 | 6 | 1 | 5 |
|---|---|---|---|---|

**Section E** (p 59)

**1** A and D (both equivalent to $3x + 5$)
B and F (both equivalent to $3x - 3$)
C and E (both equivalent to $3x + 1$)

**2** (a) $2x - 6$      (b) $5x - 16y$
(c) $12 - 2x$      (d) $2 - 10x$
(e) $10x - 27$     (f) $9x$
(g) $x + 3y$      (h) $x$
(i) $12x + 7y$    (j) $6x - 5y$

# ⑳ The right connections

This unit looks at handling bivariate data. Pupils will interpret and draw scatter diagrams, look at positive and negative correlation and draw lines of best fit on scatter graphs.

| | | |
|---|---|---|
| **T** | p 137 **A** 'The Missing Link' | Introducing scatter diagrams |
| | p 138 **B** Scatter diagrams | |
| | p 140 **C** Correlation and lines of best fit | |
| | p 142 **D** Quarters | A method for detecting correlation |
| **T** | p 144 **E** Are you fit then? | Investigation involving multivariate data |

---

**Essential**

Squared paper,
stopwatches, reaction rulers made from sheet 141
traffic cones, tape measures

**Practice booklet** pages 60 and 61

---

## 𝔸 'The Missing Link' (p 137)

**T**

This is a short introductory task to introduce scatter diagrams.
The answers are
A Phil,  B Lemon,  C Howie,  D Biffo,  E Aaron,  F Stix

## 𝔹 Scatter diagrams (p 138)

| Squared paper |
|---|

◊ You may need to explain the use of jagged lines on one or both of the axes.

**B2** The 'bleep test' is a standard fitness measure used in PE. It is based on the 'shuttle run' described in section E. A score is achieved based on how many 'runs' can be achieved in a certain period.

## C Correlation and lines of best fit (p 140)

◊ Pupils are likely to ask how they judge a line of best fit. The usual rule of thumb is to ensure that there are equal numbers of points either side of the line, but there are lots of lines that obey this but in no way fit the data. Clearly, the better the correlation, the easier it is to fit a line. In fact, unless the correlation is strong it is not worth trying.

◊ The main purpose of drawing a line is to estimate other values where this makes sense. Estimating outside the range of data is hazardous.

## D Quarters (p 142)

◊ This technique is the basis of a method used for quantifying correlation in a branch of statistics called 'exploratory data analysis'.

## E Are you fit then? (p 143)

Stopwatches, reaction ruler made from sheet 141, cones, tape measures

◊ This activity is a good way to develop skills in using and applying mathematics. Trialling has shown the exercises on this page do not require specialised supervision. With negotiation it could be done in a PE lesson. All but step-ups are accessible to pupils in wheelchairs. There are many other activities which could be used if you have equipment or facilities, for example standing jump (chalk on wall), peak flow meters used by asthmatics, grip meters etc.

◊ When the data has been collected pupils will need to record the class data on a 'spreadsheet' style grid which can be photocopied and distributed.

◊ Pupils should make their own hypotheses first and then test these using the available data. Presenting a paper is a good way for pupils to report on their findings.

This activity is considerably enhanced by using spreadsheets or graphical calculators to process the data.

## B Scatter diagrams (p 138)

**B1**

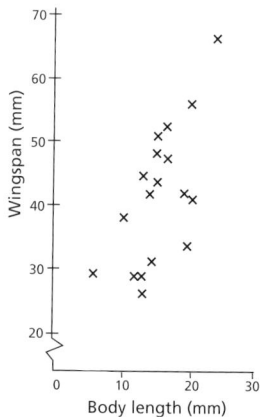

(a) Yes    (b) No    (c) Yes

**B2** (a)

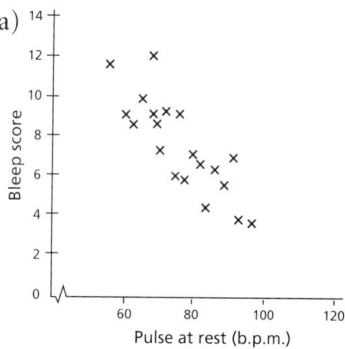

(b) Pupils with a lower rest pulse rate generally have a higher bleep score.

**B3** (a) Taller people tend to weigh more.

(b) Cars with bigger engines tend to go faster.

(c) People with longer legs generally run 100 m faster.

(d) Height and maths result are not linked.

**B4** (a) The pupil's graph showing negative correlation

(b) The pupil's graph showing zero correlation

## C Correlation and lines of best fit (p140)

**C1** About 25.5 cm

**C2** (a) Positive    (b) Positive

(c) Negative    (d) Zero

**C3** (a), (b)

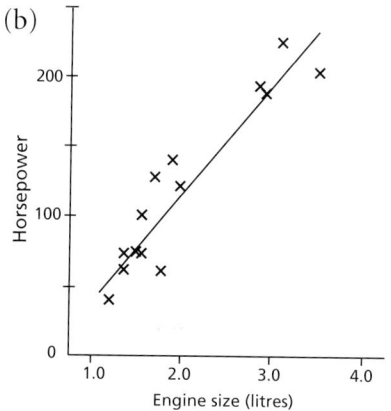

(c) The horsepower of a 2.5 litre car would be about 150.

**C4** (a)

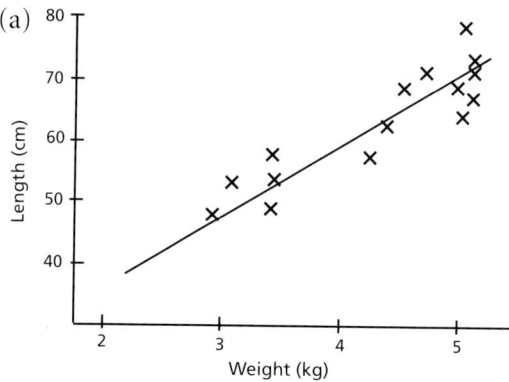

(b) (i) 60 cm    (ii) 48 cm    (iii) 56 cm
(iv) 80 cm

(c) 5.1 kg

(d) 7 kg; the estimate would not be reliable because it is outside the range of the data.

**C5** (a)

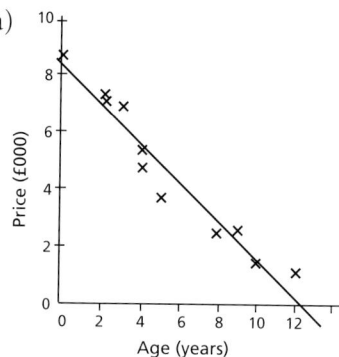

(b) About £3500

(c) About 13 years

## D Quarters (p 142)

**D1** (a) There would be more points in downward diagonal.

(b) There would be roughly equal numbers in both diagonals.

**D2**

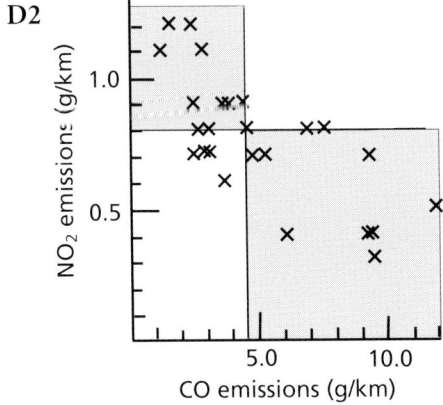

There are 4 crosses in the upward diagonal, 16 in the downward diagonal (5 on line) which suggests negative correlation.

### What progress have you made? (p 144)

**1**

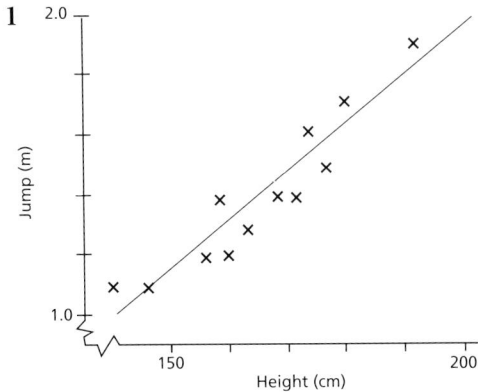

**2** Taller people generally jump further.

**3** (a) Positive

(b) The pupil's sketches of negative and zero correlation

**4** A person 2 m tall could be expected to jump about 2 m.

## Practice booklet

### Section B (p 60)

**1** (a)

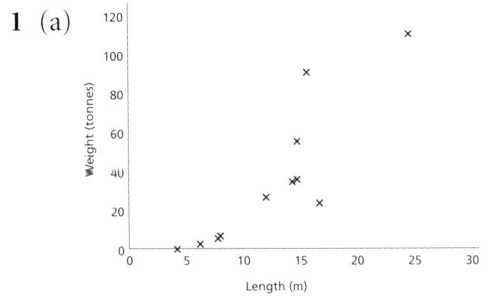

(b) Generally longer whales weigh more.

**2** (a) The Sei and the Fin

(b) 26 and 20 knots respectively

(c) This suggests these two whales are slender for their lengths and are much more streamlined.

### Sections C and D (p 61)

**1** (a) There is no correlation between weight and speed.

(b) There is no evidence to support the biologist's hypothesis.

**2** (a)
(b)

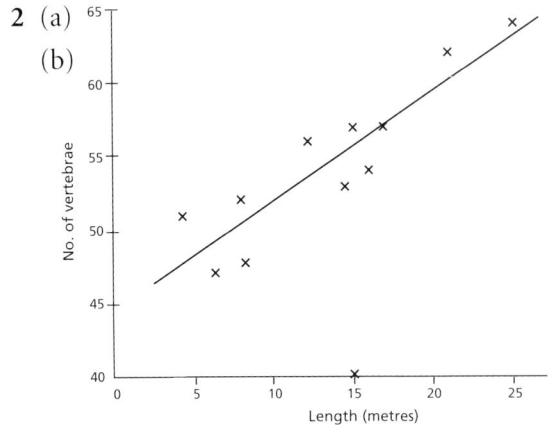

(c) Approximately 60

**3** Median length = 14.8 m
Median vertebra = 53.5

Quartering gives 10 in the leading diagonal and 2 in the other quarters. This suggests a positive correlation.

# ㉑ Triangles and polygons

This unit is mainly concerned with establishing and using angle properties. It starts with a brief recap of the construction of triangles.

---

**Essential**

Angle measurers, compasses

**Practice booklet** pages 62 to 66

---

## 𝔸 Constructing triangles (p 145)

Angle measurers, compasses

Work on constructing triangles was included in *Book 1*. Here there is the opportunity to reach an overview of the information needed to fix the size and shape of a triangle.

◊ You could ask pupils to draw accurately each of the triangles shown. For the second, they will not know enough to fix the size of the triangle, although the shape will be fixed. The last example is the most significant, as there are two possible positions for C:

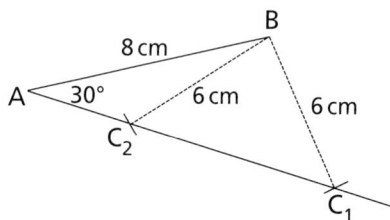

However, if the angle at A was a right angle and BC was longer than AB then the triangle would be fixed.

As you draw the work together pupils should see that the patterns SSS, SAS, ASA and RHS fix a triangle, but AAA (where length information is lacking) and SSA (the ambiguous case) do not. This leads to SSS, SAS, ASA and RHS being each a sufficient condition for the congruence of two or more triangles. Many geometrical proofs depend on showing from limited information that two triangles are congruent, from which further facts can then be deduced. For example, a challenge to follow on from section B might be 'Prove that in any rectangle each pair of opposite sides is equal: use only the fact that the angles of a rectangle are all right angles and that the angles of a triangle add up to 180° (hint: draw in a diagonal).'

## B The angles of a triangle (p 146)

◊ You may prefer to work through question B1 with the whole class. Even if you do not, it is important to help pupils to understand the difference between a practical demonstration and a proof. For example, tearing the three angles from a paper triangle and putting them against a straight edge is open to the objections that (1) you can't tell whether the angles add up exactly to 180°, (2) even if you could, you would only have demonstrated it for that one triangle.

## C Interior angles of a polygon (p 148)

◊ You could start by drawing an irregular polygon on the board or OHP and asking a volunteer to split it into triangles so that every triangle corner is a corner of the polygon. Then ask if it could be done in any other ways. (Pupils will find that the more sides there are, the more ways there are to split the polygon into triangles.) Pupils should also investigate whether the shape of the polygon makes any difference. If the pupils do not do so of their own accord, introduce a polygon with at least one reflcx angle. It is important that they realise that the generalisation includes polygons with reflex interior angles.

◊ After a while suggest that they make a table giving the number of triangles next to the number of polygon sides. Tell them to leave room for at least one more column on the right-hand side of the table.

◊ When pupils are confident that the number of triangles is always two less than the number of sides, ask how many triangles there would be in a polygon with $n$ sides, so that the expression $n - 2$ becomes well established.

They should now see why the total of the interior angles of an $n$-sided polygon is $180(n - 2)°$.

### Ⓓ **Exterior angles** (p 150)

◊ Before pupils start this section it would be effective to do the 'Sliding pieces' demonstration:

 • Draw a polygon (any polygon).
 • Extend the sides of the polygon to the edge of the paper, and mark the exterior angles.
 • Cut out all the pieces, remove the polygon and fit the pieces back again.
 • Slide the pieces (without twisting) until all the exterior angles are together.

You could use large pieces of paper prepared beforehand and fix them to a vertical surface using Blu-tak so that the pieces can easily be unstuck, moved and then re-stuck.

◊ Pupils could do the 'Pen pushing' experiment themselves first with a convex polygon and then with a concave polygon such as an arrowhead. They should be able to explain what happens at a reflex angle (the total is still 360° if the exterior angle at the reflex corner is subtracted instead of added because the pen is turning the other way).

#### Using LOGO

This activity exemplifies a common process in the development of mathematics: take an idea or procedure which is defined for a restricted type of number (positive integers in this case) and see what happens if the restriction is removed but all the other rules still followed. (An analogous example which will come much later is the extension of powers to include negative numbers and fractions.)

This is how LOGO draws a 'regular polygon with $2\frac{1}{2}$ sides':

### Ⓐ **Constructing triangles** (p 145)

**A1** (a) The pupil's triangle
  (b) AC = 5.7 cm, BC = 4.8 cm, angle C = 90°

**A2** (a) The pupil's triangle
  (b) Angle P = 62°, angle Q = 84°, angle R = 34°

**A3** (a) The pupil's two triangles
  (b) XZ = 13.1 cm or 5.8 cm

**A4** (a) Yes
  (b) 3.5, 5.5, 7.5
     3.5, 7.5, 9.5
     3.5, 9.5, 11.5
     3.5, 11.5, 13.5
     5.5, 7.5, 9.5
     5.5, 7.5, 11.5
     5.5, 9.5, 11.5
     5.5, 9.5, 13.5
     5.5, 11.5, 13.5
     7.5, 9.5, 11.5
     7.5, 9.5, 13.5
     7.5, 11.5, 13.5
     9.5, 11.5, 13.5

## B The angles of a triangle (p 146)

**B1** (a) They are corresponding angles.

(b) They are alternate angles.

(c) Angles ECD, BCE and $c$ are supplementary angles on a straight line, so $a + b + c = 180°$.

**B2** (a) $v = 45°$ (exterior angle of triangle)

(b) $w = 76°$ (exterior angle of triangle, base angles of isosceles triangle)

(c) $x = 28°$ (base angles of isosceles triangle, exterior angle of triangle)

(d) $z = 97°$ (base angles of isosceles triangle, exterior angle of triangle)

(e) $y = 60°$ (alternate angles, exterior angle of triangle)

**B3** (a) 4 triangles with angle sum $180°$
$4 \times 180 = 720°$

(b) $360°$

(c) The pupil's explanation

**B4** (a) $p = 21°$ (angle sum of quad = $360°$)

(b) $q = 78°$ (angles on straight line, angle sum of quad = $360°$)

(c) $s = 66°$ (exterior angle of triangle, angle of sum of quad = $360°$)

**B5** (a) $x = 25°$      (b) $y = 45°$

## C Interior angles of a polygon (p 148)

**C1** (a) $1800°$ ($10 \times 180°$)

(b) $18\,000°$ ($100 \times 180°$)

**C2** (a) $a = 95°$      (b) $b = 100°$

(c) $c + d = 128°$, $d = 48°$, $c = 80°$

(d) $e = 60°$ (angles on straight line)
$f = 110°$ (angle sum of quad)
$g = 70°$ (angles on straight line)
$h = 130°$ (angle sum of pentagon)
$i = 60°$ (angle sum of triangle)

**C3** (a) $720°$

(b) The reflex angle is $270°$, not $90°$.

**C4** $57°$

**C5** $2(p + q) = 540° - 140°$
$2(p + q) = 400°$
$\quad p + q = 200°$

**C6** (a) to (d) The pupil's investigation

(e) Total of interior angles = $(180n - 360)°$

(f) $(180n - 360)° = 180(n - 2)°$

**C7** 15

**C8** (a) $108°$      (b) $135°$

(c) $144°$      (d) $150°$

**C9** $128\frac{4}{7}°$

**C10** 18 sides

## D Exterior angles (p 150)

**D1** (a) $(n - 2)180°$      (b) $360°$

(c) $(n - 2)180° + 360° =$
$180n° - 360° + 360° = 180n°$

**D2** There are $n$ corners.
Exterior + interior angles at each corner = $180°$.
Therefore grand total = $(180 \times n)°$

**D3** $a = 113°$,  $b = 55°$,  $c = 53°$

**D4** $x = 48°$

**D5** (a) $72°$  (b) $45°$  (c) $40°$  (d) $36°$

**D6** The interior angle is
$180°$ – the exterior angle.

(a) $108°$      (b) $135°$

(c) $140°$      (d) $144°$

**D7** (a) $18°$      (b) $162°$      (c) $156°$

**D8** They are equal. They are both obtained by dividing $360°$ by the number of sides (triangles).

**D9** $x = 72°$

**D10** The interior angles of polygons meeting at P are $90°$, $135°$, $135°$.
$90 + 135 + 135 = 360°$

**D11** Interior angle of A
$$= 360 - 135 - 60 = 165°$$
$$\frac{180(n-2)}{n} = 165$$
$$180n - 360 = 165n$$
$$n = 24$$
So polygon A has 24 sides.

**D12** (a) $\dfrac{360}{n}$      (b) $\dfrac{180(n-2)}{n}$

**D13** (a) 12 sides      (b) 18 sides
     (c) 15 sides      (d) 42 sides

### What progress have you made? (p 153)

1 (a) The pupil's triangle
  (b) AC = 3.5 cm or 12.8 cm

2 $x = 44°$, $y = 82°$

3 2160°

4 (a) 20°      (b) 160°

5 (a) 12°      (b) 30 sides

## Practice booklet

### Section A (p 62)

1 (a) The pupil's triangle
  (b) 9.6 cm

2 (a) The pupil's two triangles
  (b) 7.8 cm, 4.7 cm

### Section B (p 62)

1 $a = 55°$     $115° - 60°$
  $b = 59°$     $38° + 21°$
  $c = 62°$     $180° - 2 \times 59°$
  $d = 64°$     $32° \times 2$
  $e = 112°$    $48° + 64°$

2 $x = 40°$

3 $r = 43°$     $360° - 90° - 122° - 105°$
  $s = 135°$    $60° + 45° + 30°$
  $t = 133°$    alternate angles
  $u = 54°$     $101° - (180° - 133°)$

4 60° and 120°

### Section C (p 63)

1 (a) 720°      (b) 1080°

2 The pupil's reasoning with
  $a = 140°$, $b = 29°$, $c = 125°$, $d = 80°$

3 13 sides

4 150°

5 170°

6 125°

7 110° (the two unmarked angles total 180°)

8 $x = 90°$, $y = 60°$

### Section D (p 65)

1 The pupil's explanation;
  $a = 145°$, $b = 34°$, $c = 70°$, $x = 30°$

2 155°

3 (a) 60°      (b) $51\frac{3}{7}°$

4 9 sides

5 (a) 45° (b) 30° (c) 18° (d) 1°

6 (a) 108°    (b) 135°    (c) 150°
  (d) 171°    (e) 178°

7 $a = 36°$   half of $(180° - 108°)$
  $b = 36°$   $108° - 2 \times 36°$
  $c = 72°$   half of $(180° - 36°)$
  $d = 30°$   half of $(180° - 120°)$
  $e = 60°$   half of 120°, symmetry
  $f = 90°$   $120° - 30°$
  $g = 135°$ $180° - 45°$
  $h = 67\frac{1}{2}°$ half of $(180° - 45°)$
  $i = 22\frac{1}{2}°$ half of $(180° - 135°)$
  $j = 90°$   $67\frac{1}{2}° + 22\frac{1}{2}°$

8 $x = 90°$   $180° - 2 \times 45°$
  $y = 45°$   $360° - 2 \times 45° - (180° + 45°)$

# 22 Moving around

| **Essential** | **Optional** |
|---|---|
| Sheet 207 | Measuring tape, watches |
| **Practice booklet** pages 67 and 68 | |

## A How fast? (p 154)

> Sheet 207
> Optional: measuring tape, watches

◊ An opening discussion should help you find out what pupils know about speed. Ask them to estimate the speeds shown in the pictures, perhaps first putting them in order. Note the units they use – what do they mean by, for example, '5 miles an hour'?

**How fast do you walk?**

◊ Discuss how walking speeds could be measured.

A possible way is to choose, for example, a hall or tennis court where pupils can repeatedly walk across for a measured time, say 2 minutes. They measure the distance across, count how many times they have walked it and measure any extra partial crossing.

The result in metres per minute can be converted to miles per hour using the conversion graph.

**Using a map**

◊ It is better if the work can be based on a local map.

◊ In hilly country, 'Naismith's rule' can be used to help estimate journey times. It is 20 minutes per mile plus 1 hour for every 2000 feet climbed.

## B Distance, time and speed (p 156)

T

◊ The 'pigeons' map is for class discussion. You could divide the class into groups and ask each group to put the pigeons in order of speed.

◊ Emphasise that all speeds calculated are **average** speeds.

◊ The train and air journey data could be used together as a class or by pupils working in pairs or groups. The average speeds (to the nearest 0.1 m.p.h) are:

### Train journeys

Perth 75, Doncaster 104, Walton on the Naze 46.7, Shoeburyness 40, Hastings 42, Gatwick Airport 52, Weymouth 57.2, Penzance 61, Swansea, 63.7, Holyhead 66, Glasgow 80.2, Sheffield 66.

### Air journeys

Copenhagen 348.6, Moscow 451.4, Rome 420, Paris 304, Cardiff 186.7, Orlando 456.2, Toronto 426.3, Derry 184.

## C Constant speeds (p 158)

## D Calculating times (p 159)

## E Time on a calculator (p 160)

## F Mixed questions (p 161)

### A How fast? (p 154)

**A1** (a)  About 1300 metres

(b)  About 10–15 minutes

**A2**  About 600 metres

**A3**  The distance is about 2000 metres. It would take about 25–30 minutes.

**A4**  The pupil's instructions

### C Constant speeds (p 158)

**C1**  21 m

**C2** (a)  12 m    (b)  20 m    (c)  120 m

(d)  240 m    (e)  1200 m

**C3**  480 m

**C4** (a)  750 m    (b)  2000 m

**C5**  144 m

**C6**  15 m

**C7**  150 miles

**C8**  135 miles

**C9**  5750 km

**C10** (a)  6 hours    (b)  15 hours

**C11** (a)  4 hours    (b)  7 hours    (c)  $2\frac{1}{2}$ hours

**C12** (a)  180 miles    (b)  90 miles

(c)  6 miles

**C13** 48 m.p.h.

**C14** (a) 36 m.p.h.     (b) 40 m.p.h.
     (c) 75 m.p.h.     (d) 66 m.p.h.

**C15** (a) BC       (b) AB
     The speeds are
     AB 32 m.p.h., BC 39 m.p.h.,
     CD 33 m.p.h., DE 35 m.p.h.

### D Calculating times (p 159)

**D1** (a) 3 hours     (b) $2\frac{1}{2}$ hours
     (c) $\frac{1}{2}$ hour

**D2** (a) 5 hours     (b) $3\frac{3}{4}$ hours
     (c) $\frac{1}{4}$ hour

**D3** The pupil's explanation
     Time taken = $900 \div 200 = 4\frac{1}{2}$ hours

**D4** $2\frac{1}{2}$ hours

**D5** (a) 8 hours     (b) 4 hours
     (c) 1 hour      (d) 6 hours

**D6** 3.25 hours is $3\frac{1}{4}$ hours,
     or 3 hours 15 minutes.

### E Time on a calculator (p 160)

**E1** (a) 15 minutes   (b) 12 minutes
     (c) 6 minutes    (d) 27 minutes
     (e) 57 minutes

**E2** (a) 0.78 hour   (b) 0.67 hour
     (c) 0.35 hour   (d) 0.13 hour
     (e) 0.95 hour

**E3** (a) 2 hours 19 minutes
     (b) 4.28 hours

**E4** 2 hours 17 minutes

**E5** 09:11

**E6** Sandra's journey will take 2.73 hours, or
     2 hours 44 minutes (to the nearest
     minute). She will be 14 minutes late.

### F Mixed questions (p 161)

**F1** (a) (i) Debbie
        (ii) Alex
        (iii) Fergus
        (iv) Fergus
        (v) Crystal
     (b) Alex and Ella

**F2** 7.6 m/s

**F3** 1 hour 22 minutes

**F4** 48 minutes

**\*F5** 21 m.p.h.

### What progress have you made (p 162)

**1** (a) 3 m/s      (b) 37 m.p.h.
   (c) 240 km/h

**2** (a) 125 m      (b) 135 miles
   (c) 1300 km

**3** (a) 6 hours
   (b) $5\frac{1}{2}$ hours
   (c) $2\frac{1}{4}$ hours

**4** 0.37 hour

**5** (a) 2 hours 20 minutes
   (b) 50 minutes

### Practice booklet

### Section B (p 67)

**1** (a)

| Queenstown | Nelson | Kaikoura | Greymouth | Christchurch |
|---|---|---|---|---|
| | | | | 68.0 |
| | | | 67.6 | 75.6 |
| | | 76.6 | 78.9 | 69.5 |
| | 66.7 | 71.2 | 61.6 | 69.6 |

Average speed in km/h

     (b) Greymouth to Nelson
     (c) Greymouth to Queenstown

**Section C** (p 67)

 1 (a)  375 miles       (b)  62.5 miles

 2 (a)  150 miles       (b)  135 miles

 3 (a)  44 km           (b)  6 km

 4  42 miles

 5 (a)  52.5 m.p.h.     (b)  36 m.p.h.
   (c)  32 mp.h.        (d)  48 mp.h.

**Sections D and E** (p 68)

 1 (a)  3 hours         (b)  5 hours
   (c)  $1\frac{1}{2}$ hours

 2 (a)  12 hours        (b)  9 hours
   (c)  4 hours

 3 (a)  3 hours 8 minutes
   (b)  2.87 hours

 4  1 hour 49 minutes

# ㉓ Substitution

The emphasis in this unit is on the comprehension of the algebraic shorthand which a formula represents, on care over use of units and on the accurate substitution of numbers with and without the help of a calculator.

## Units

Often formulas involve quantities such as area, length, speed etc. When scientists and engineers use such formulas, the letters usually stand for 'numbers-with-units'. Mathematicians more usually use letters simply to stand for pure numbers.

Pupils are likely to meet both conventions at about the same stage in their school careers. You may therefore think it important to talk about the difference between the scientist's style and the mathematician's. Each is simply a matter of convention – neither is more correct than the other. A dialogue between mathematics teachers and science teachers over this matter, and over other conventions such as, for example, plotting and labelling graphs, may help to avoid confusion on the pupils' part.

Here is a simple example of finding an area written out in the two styles.

| | |
|---|---|
| *The scientist's style:* | *The mathematician's style:* |
| *Let L be the length.* | *Let the length be L cm.* |
| *Let B be the breadth.* | *Let the breadth be B cm.* |
| *Let A be the area.* | *Let the area be A cm².* |
| *Then L = 3 cm* | *Then L = 3* |
| *and B = 2 cm* | *and B = 2* |
| *A = LB = 3 cm × 2 cm* | *A = LB = 3 × 2 = 6.* |
| *= 6 cm².* | *So the area is 6 cm².* |

## Accuracy

Often there is another difference between the scientist's and the mathematician's style: the scientist thinks in terms of measurements and normally gives data to whatever accuracy the means of measurement permits. Thus, the scientist would almost certainly not write as above *Let L = 3 cm* because that would imply an accuracy of measurement such that $L$ lies between 1.5 and 2.5 cm. For a measurement with a ruler, for instance, the scientist's version might, more likely, be *Let L = 3.0 cm*.

At this stage, it is probably best for pupils not to be too distracted by considerations of accuracy. So the numbers used in problems in the exercises lead to results which will not demand attention to the problem of the accuracy of the answer. In a few cases only a particular degree of accuracy is asked for.

---

**Practice booklet** pages 69 and 70

---

## Ⓐ **Letters for numbers** (p 163)

T

◊ Finding the areas of the three rectangles shown can be used to bring out the points in the 'When using formulas' box.
The areas of the rectangles are $3\,cm^2$, $3.6\,m^2$ and $1.2\,km^2$.

## Ⓑ **Further formulas** (p 165)

T

This section contains a large number of questions. It is not intended that these are all completed at one sitting.

◊ Pupils will need to be aware of how their calculator works. The examples at the start of the section may be used to start group or class discussion about this.

◊ You may need to revise rounding to significant figures.

## Ⓐ **Letters for numbers** (p 163)

**A1** (a) 640 m     (b) 210 cm
(c) 86 mm or 8.6 cm
(d) 390 cm or 3.9 m

**A2** (a) $24\,000\,m^2$     (b) $2600\,cm^2$
(c) $450\,mm^2$ or $4.5\,cm^2$
(d) $0.9\,m^2$ or $9000\,cm^2$

**A3** (a) $6\,m^2$    (b) 16.6 m    (c) £43.50

**A4** (a) $T = 3$     (b) 4 hours
(c) $T$ gets smaller. (d) $T$ gets larger.
(e) She will never get there as she is not moving at all.

**A5** (a) 60 cm
(b) 50 cm; at 2 a.m.
(c) 37.5 cm
(d) After 12 hours, at noon
(e) $d = {}^-5$. It doesn't make sense – you can't have a negative depth.

**A6** (a) The pupil's validation of the surface area formula
(b) (i) Vol $3000\,cm^3$, SA $1340\,cm^2$
(ii) Vol $120\,m^3$, SA $166\,m^2$

**A7** (a) $20\,cm^3$
(b) It is not sensible to use the formula, since the area of the 'edges' of the paper is tiny, so can be ignored.

## Ⓑ **Further formulas** (p 165)

**B1** (a) 30     (b) 15     (c) 1
(d) 4.5     (e) 180

**B2** (a) 0.23     (b) 5.22     (c) 20.39
(d) 0.58     (e) 20.18

**B3** (a) $^-22$     (b) $^-0.2$     (c) 16
(d) 360     (e) 20

**B4** (a) 30     (b) 4.2     (c) 180
(d) 320     (e) 77

**B5** (a) 3.375     (b) 12.25     (c) 0.75
(d) 2.25     (e) 7.75

**B6** (a) $2\frac{1}{2}$     (b) 5     (c) $11\frac{2}{3}$
(d) $17\frac{1}{2}$     (e) 36

**B7** (a) $2800\,cm^2$ (b) $0.92\,m^2$ or $9200\,cm^2$

**B8** (a) $8.4\,cm^2$    (b) $0.51\,km^2$

**B9** (a) Volume = $ab^2$ cm$^3$

(b) 48 000 cm$^3$    (c) 1 000 000 cm$^3$

(d) (i) 1 010 000 cm$^3$
    (ii) 1 020 100 cm$^3$

**B10** (a) 12     (b) 18     (c) 36

**B11** 3, 1, 2    3, 2, 1    4, 1, 3
     4, 3, 1    5, 2, 3    5, 3, 2

**B12** 2700 kg

**B13** $c = 1.05$

**B14** (a)

| $t$ | 0 | 1 | 2 | 3 | 4 |
|-----|-----|-----|-----|-----|-----|
| $h$ | 80 | 75 | 60 | 35 | 0 |

(b) It speeds up.    (c) After 4 seconds

**B15** (a)

| $S$ | 0 | 10 | 20 | 30 | 40 | 50 | 60 | 70 |
|-----|-----|-----|-----|-----|-----|-----|-----|-----|
| $D$ | 0 | 5 | 13.3 | 25 | 40 | 58.3 | 80 | 105 |

(b) The pupil's graph
(c) 45 or 46 m.p.h.

**B16** (a) 1130
(b) (i) 1730    (ii) 820
(c) 1120 cm$^2$

**B17** (a) 6 mg      (b) 24 mg

*$\star$**B18** (a) 1.996 seconds
(b) (i) 1800    (ii) 1805
(c) 0.996 metres

### What progress have you made? (p 170)

**1** (a) 2.4 m    (b) 3.2 m
(c) 3.5 m    (d) 2.1 m

**2** (a) 14    (b) 2    (c) 4
(d) 14    (e) 18    (f) 20

**3** (a) 0.12    (b) ⁻2
(c) ⁻0.9    (d) 1.944

**4** (a) 182    (b) 217.5 m

### Practice booklet
### Section A (p 69)

**1** (a) 7 cm$^2$      (b) 400 mm$^2$ or 4 cm$^2$

**2** (a) 800      (b) 225

### Section B (p 69)

**1** (a) 45   (b) 15   (c) 5   (d) 200

**2** (a) 8.5    (b) 2    (c) 4
(d) 12    (e) 1

**3** (a) 36    (b) 48    (c) 144

**4** (a) 12    (b) 36    (c) 87

**5** 3.6

**6** (a)

| $t$ | 0 | 1 | 2 | 3 | 4 | 5 | 6 | 7 | 8 |
|-----|-----|-----|-----|-----|-----|-----|-----|-----|-----|
| $h$ | 0 | 45 | 80 | 105 | 120 | 125 | 120 | 105 | 80 |

(b)

| $t$ | 0 | 1 | 2 | 3 | 4 | 5 | 6 | 7 | 8 |
|-----|-----|-----|-----|-----|-----|-----|-----|-----|-----|
| $v$ | 50 | 40 | 30 | 20 | 10 | 0 | ⁻10 | ⁻20 | ⁻30 |

(c) The pupil's interpretation of tables (a) and (b). They might include in their answer that the maximum height of the stone is 125 m after 5 seconds. At 5 seconds the speed is 0 and the stone then descends which is why the values for the speed are negative.

# ㉔ Locus

| Essential | Optional |
|---|---|
| Sheet 208 | Drawing pins, string |
| Compasses | |
| **Practice booklet** pages 71 and 72 | |

## ⒜ **Place the points** (p 171)

**T**

During the initial discussion pupils could participate in placing points on the board or on an OHP. Since the idea of a locus as 'the set of all points that …' is sometimes found difficult, activities with counters can sometimes provide a good start. They also give confidence in dealing with questions where points have to fulfil more than one condition. This is a good opportunity to practise skills of estimating distances.

'Without measuring, try to place an orange counter 20 cm from the white counter. Now check with a ruler and correct the orange counter's position. Do the same with your other orange counters.'

'Place a coin so that it is less than 20 cm from the middle counter.'

'Suppose I wanted to use red paint to show all the possible places you could put a coin less than 20 cm from the middle counter. Where would I paint?

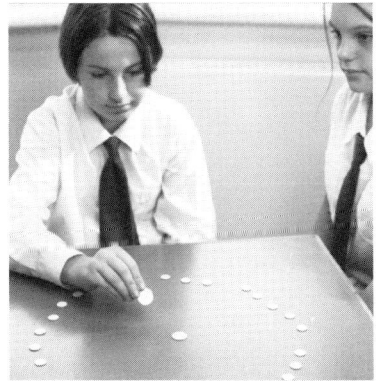

'Place a coin so it is more than 20 cm from the middle counter but less than 30 cm from it.'

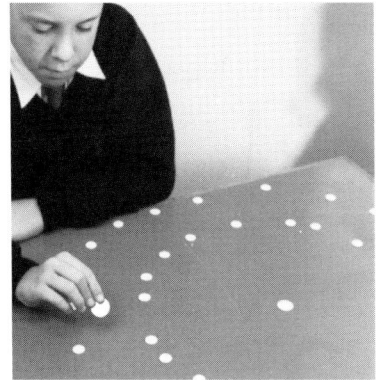

'Place each green counter so that it is 15 cm from the short edge of the table.'

'Show me a point that is more than 15 cm from the short edge of the table and is less than 25 cm from the corner.'

'Suppose I wanted to use yellow paint to show all the possible points that meet those two conditions, where would I paint?

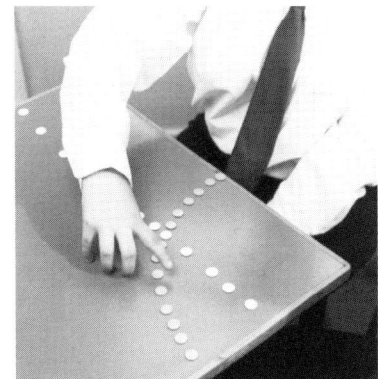

## B Regions and boundaries (p 171)

> Sheet 208, compasses

◊ This covers essentially the same ground as in section A, but is set in a context.

## C Sets of points (p 172)

> Compasses
> Optional (for question C11): drawing pins, string

## D Moving points (p 175)

## E Loci in three dimensions (p 176)

◊ It is better if you demonstrate practically. For example, hold out a pencil and use its end as a fixed point. Ask a pupil to put the tip of their finger one metre from the fixed point. Where could they put their finger?

### B Regions and boundaries (p 171)

**B1** (a) The pupil's answer on sheet 208

(b) A circle, centre A, radius 5 km

**B2** The boundary is a circle, centre A, radius 7 km.

**B3** There is no unique solution.

**B4** (a)–(c) The pupil's answer on sheet 208

(d) The boundary is the perpendicular bisector of BC.

### C Sets of points (p 172)

**C1**

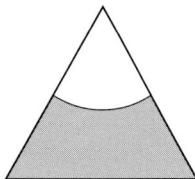

**C2**

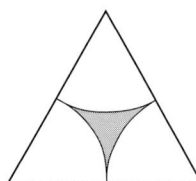

**C3**

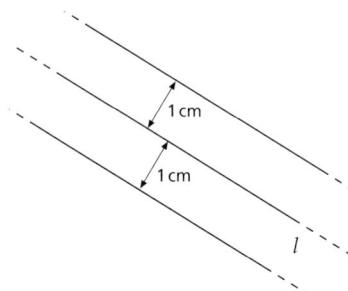

**C4**

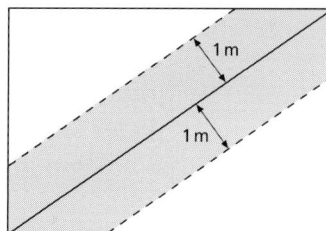

**C5**

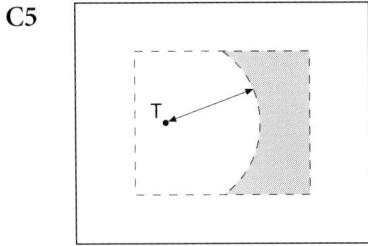

**C6**

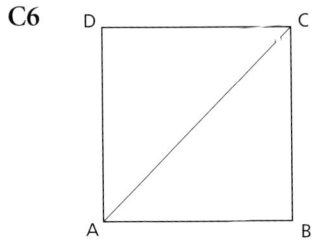

**C7**

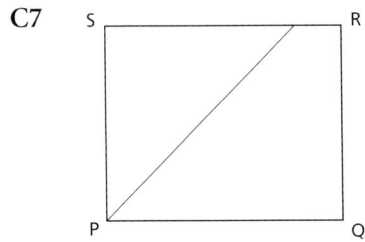

**C8**

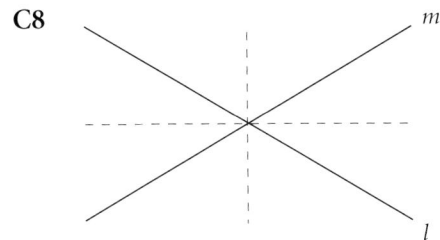

The locus consists of the two bisectors of the angles *l* and *m*.

**C9**

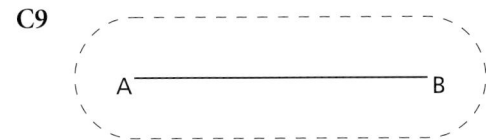

**C10**

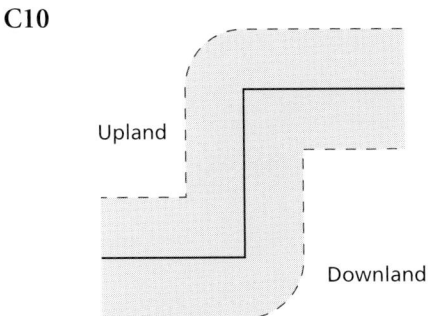

**C11**

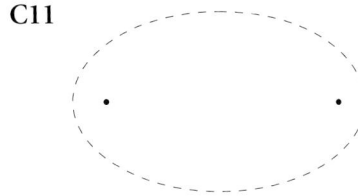

The locus is an ellipse.

**\*C12**

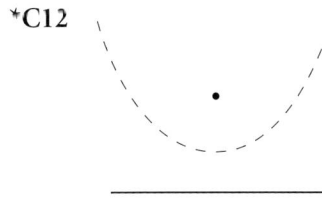

The locus is a parabola.

▷ **Moving points** (p 175)

**D1**

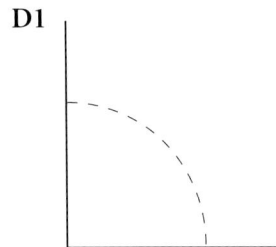

The locus is an arc of a circle.

**D2**

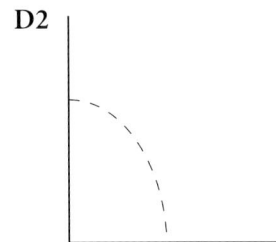

The locus is part of an ellipse.

**D3**

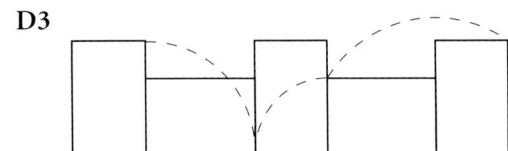

**D4**

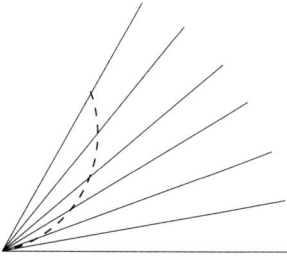

**\*D5**

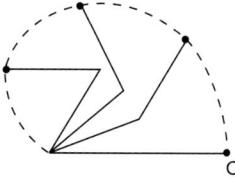

## \*E Loci in three dimensions (p 176)

**E1** A sphere, centre A, radius 1 metre

**E2** The plane that bisects AB and is perpendicular to AB

**E3** Two parallel planes on either side of *p*, distance 10 cm from *p*

**E4** A cylinder of radius 10 cm whose axis is *l*

**E5** (a) A circle     (b) A cone

### What progress have you made? (p 176)

**1**

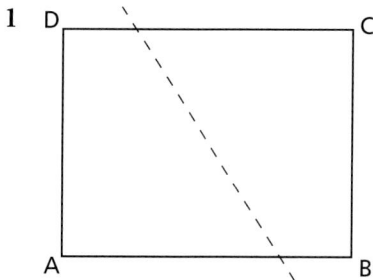

The locus is the perpendicular bisector of AC.

**2**

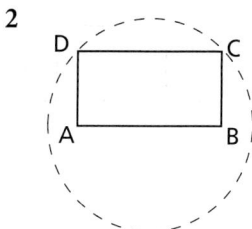

The locus is a circle.

## Practice booklet

### Section C (p 71)

**1**

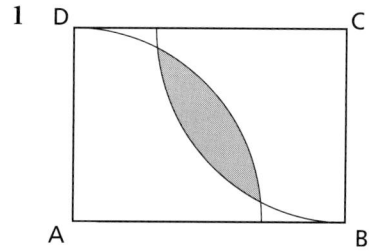

**2** (a)

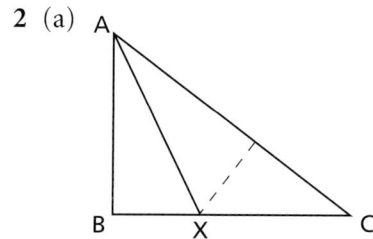

   (b) BX = 3 cm

### Section D (p 72)

**1**

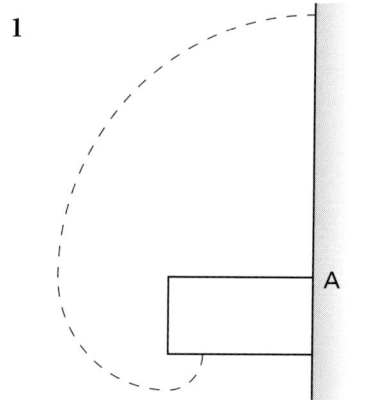

**2**

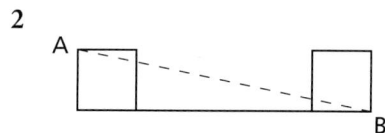

The locus is the straight line AB.

# Ⓩ Distributions

This unit revises earlier work on median, mode, mean and range, and introduces stem-and-leaf tables and frequency polygons. It also covers calculating an estimate of the mean from a grouped frequency table.

Practice booklet pages 73 to 78

## Ⓐ Median, range and mode (p 177)

◊ Questions A7–A10 informally introduce the use of an assumed mean.

## Ⓑ Stem-and-leaf tables (p 179)

◊ Measuring pulse rates provides suitable data for a stem-and-leaf table, and the data is relatively easy to collect. When pupils are sure that they have found their pulse, you can start and finish a one-minute interval. The numbers of beats per minute are recorded on a class table similar to this:

```
4 | 7              ←——————  This shows a pulse rate
5 | 6 3 8 2                  of 47 beats per minute.
6 | 3 7 1 0 5
7 | 2 6 3 2
```

When all the data has been entered, the table is rewritten with the units digits in order of size:

```
4 | 7
5 | 2 3 6 8
6 | 0 1 3 5 7
7 | 2 2 3 6
```

Working on squared paper helps to keep the digits in line.

◊ Discuss how to describe the modal group (60–69 in the example above) and how to find the median and range from a stem-and-leaf table.

## Ⓒ Mean (p 180)

This section revises earlier work on finding the mean from an ungrouped frequency table.

## D Discrete and continuous data (p 182)

◊ There is no need to make too much of the distinction between discrete and continuous data. In practice, continuous measurements are rounded off to a certain degree of accuracy, and in that form are actually discrete.

◊ Throughout this unit a boundary value is put into the upper interval. The problem about where to include a boundary value can be solved by using inequalities to describe intervals, for example $2 \leq w < 3$. You may wish to introduce this notation now. (It is dealt with in a later unit.)

**D2** (d) Even though half-sizes exist, there are still jumps between sizes.

## E Frequency polygons (p 184)

## F Mean of a grouped frequency distribution (p 185)

## G Summarising and comparing data (p 187)

◊ The mode, median or mean may or may not be a useful measure of average: it depends on the data. Question G2 gives an example where none of the three gives a good 'average' because it is debatable whether there is such a thing as a 'typical' value for this set of data.

### A Median, range and mode (p 177)

**A1** (a) 46 kg      (b) 20 kg

**A2** Median 141 cm, range 34 cm

**A3** 3

**A4** (a) 39      (b) 2

**A5** 3

**A6** A3 : 3    A4 : 1    A5 : 3

**A7** (a) 6 cm      (b) 166 cm

**A8** 35.75

**A9** (a) 10 years 8 months
     (b) 1 year 5 months

**A10** 2 hours 55 minutes

**A11** (a) 792 kg      (b) 72.5 kg

**A12** 75 kg

**A13** 74

### B Stem-and-leaf tables (p 179)

**B1** (a) 48    (b) 27    (c) 73    (d) 70–79

**B2** Paper 1

```
2 | 8
3 | 8 9
4 | 1 3 4 5 9 9
5 | 0 4 4 4 5 8
6 | 0 1 2 3 3 6 6 6
7 | 0 0 2 9
8 | 0 2 2
```

Paper 2

```
2 | 8 9
3 | 4 5 7 8 8 9
4 | 0 1 2 4 4 7 8 9
5 | 0 1 2 4 5 6 8
6 | 1 2 3 4 9
7 | 0
8 | 3
```

Paper 2 appears harder. The median marks are 59 on paper 1 and 48.5 on paper 2. Most marks on paper 1 are in the 40–70 range, but on paper 2 they are in the 30–60 range.

**B3** (a)

```
1. | 799
2. | 3556788
3. | 01233478
4. | 01
```

(b) 2.4 kg    (c) 2.9 kg

## C Mean (p 180)

**C1** (a) Add up the frequencies: 20

(b) Frequency for '3 people' in table: 7

(c) Multiply $3 \times 7$: 21

(d) Add
$(1 \times 6) + (2 \times 4) + (3 \times 7) +$
$(4 \times 2) + (5 \times 1) = 48$

(e) Divide 48 by 20 = 2.4

**C2** (a) 6    (b) 21    (c) 15

(d) 53    (e) 2.5

**C3** (a) 34    (b) 124    (c) 3.6

**C4** (a) 50    (b) 1725    (c) 34.5

**C5** (a) 28    (b) 79    (c) 2.8

## D Discrete and continuous data (p 182)

**D1** (a) About 3.7–3.8 kg

(b) Because you cannot see from the graph what the smallest and largest heights are, only the intervals they are in

**D2** (a) Continuous    (b) Continuous

(c) Discrete    (d) Discrete

**D3** (a) He may be right. The graph shows that the longest journey is in the interval 20–25 minutes.

(b) We do not know the shortest and longest journey times.

(c) 5–10 minutes

(d) 40%    (e) 15%

**D3** (a)

(b)

(c)

(d)

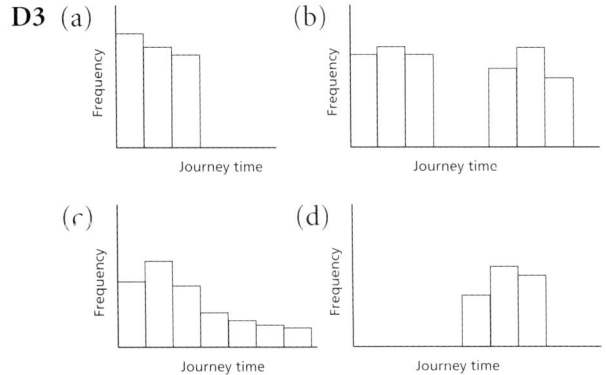

## E Frequency polygons (p 184)

**E1** (a) 30    (b) 5–10 minutes

(c) 15–20 minutes

(d) Journey times for B are longer on the whole.

**E2**

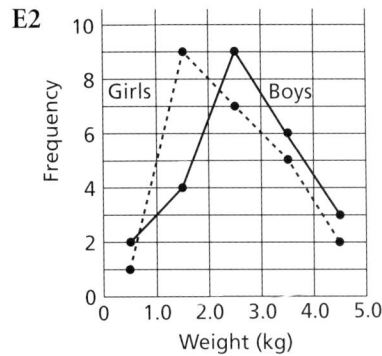

## F Mean of a grouped frequency distribution (p 185)

**F1** (a)

| Weight in kg | Mid-interval value | Frequency | Mid-interval value × frequency |
|---|---|---|---|
| 20–30 | 25 | 6 | 150 |
| 30–40 | 35 | 5 | 175 |
| 40–50 | 45 | 3 | 135 |
| 50–60 | 55 | 2 | 110 |
| Totals | | 16 children | 570 kg |

(b) 35.6 kg

**F2** 50.8 kg

**F3** 98.7 g

**F4** (a) 28.1      (b) 26–30

**F5** (a) The longest journey time is in the interval 30–35 minutes.

(b) 20–25 minutes

(c) 64%    (d) 80%    (e) 32%

(f) 20.3 minutes

Ⓖ **Summarising and comparing data** (p 187)

**G1** (a) £17k. It does not give a good idea of a typical salary.

(b) Median £11k, mode £9k. The median gives a better idea of a typical salary.

**G2** (a) Mean £21.5k, median £14k, mode £12k

(b) None of them give a good idea of a 'typical' salary. (There is no 'typical' salary in this company.)

**G3** Company A: mean £16k, range £20k
Company B: mean £16k, range £12k
The mean salary is the same for both companies.
The salaries in company B are less spread out (or closer together).

*G4 (a) True. With 9 employees and a mean of £20k, the total must be $9 \times £20k$.

(b) May be true. For example the salaries could be 10k, 20k, 20k, 20k 20k, 20k, 20k, 20k, 30k

(c) May be true. For example, the salaries could be 10k, 10k, 10k, 10k, 10k, 10k, 10k, 10k, 100k

(d) Must be false. The lowest is 10k, so the highest would be 120k. The total has to be 180k and this leaves only 50k in total for the remaining 6 employees.

**What progress have you made?** (p 188)

1 (a) 25      (b) 43      (c) 43

2

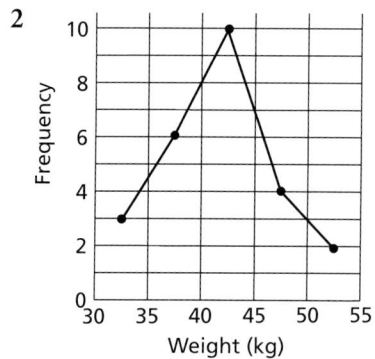

3 41.7 kg

**Practice booklet**

**Section A** (p 73)

1 (a) 18 g      (b) 4 g

2 Median 51 mm, range 12 mm

3 (a) 5      (b) 6

4 (a) 25   (b) 5   (c) 7   (d) 6

5 Zoe's mice litter size has a smaller range. Sam's mode is higher. Both have the same median.

**Section B** (p 74)

1 (a) 51 minutes     (b) 27
(c) 28 minutes     (d) 20–29 minutes

2 (a) (i) Soil A

```
2 | 1 4 5 6
3 | 0 2 2 4 6 8 9
4 | 0 1 1 3 5 7 7 8
5 | 1 6 7 9
6 | 0 2
```

Soil B

```
2 | 6 8 9 9
3 | 0 3 3 5 5
4 | 1 2 2 4 6 6 7 8 8 8
5 | 0 0 1 1 3 4
```

(ii) A: 41 cm    B: 28 cm

(iii) A: 41 cm    B: 44 cm

(iv) A: 40–49 cm    B: 40–49 cm

(b) Plants grow more consistently in soil B (smaller range in B).
The median is higher in B, although some plants grow much taller in A.

**3** (a)
```
1. | 5 8 9
2. | 0 2 4 6 6 7 8 9
3. | 1 1 2 4 4 7 9
4. | 0 1 2 2 3 4
```

(b) 2.9 kg    (c) 3.1 kg

## Section C (p 75)

**1** (a) 61.6%    (b) 61.2%

**2** (a) Girls: mean 240    range 217
Boys: mean 215.6    range 252

(b) On average the boys write fewer words per page. Girls' writing tends to be more similar in size (or similar).

(c) 226.88

**3** (a) 30    (b) 1494    (c) 49.8

**4** (a) 32    (b) 65    (c) 2.0

## Sections D and E (p 76)

**1** (a)

| No. of pens | Sasha | Hadeel |
|---|---|---|
| 1–5 | 9 | 4 |
| 6–10 | 7 | 7 |
| 11–15 | 5 | 6 |
| 16–20 | 2 | 6 |

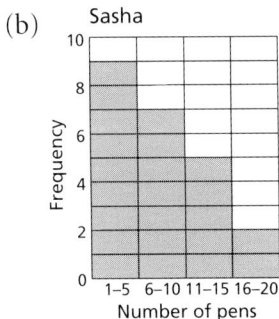

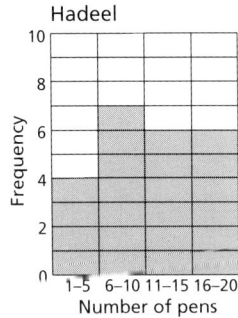

(b)

Sasha

(c) 1–5    (d) 6–10

**2** (a)

| | Boys | Girls |
|---|---|---|
| 1.20–1.30 | 1 | 3 |
| 1.30–1.40 | 5 | 6 |
| 1.40–1.50 | 5 | 4 |
| 1.50–1.60 | 4 | 2 |

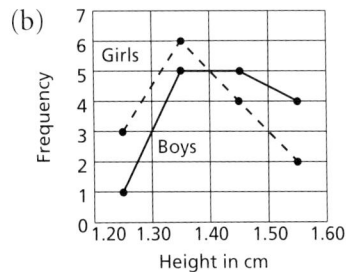

(b)

(c) Overall the boys seem to be taller than the girls.

**3** (a) 30–40    (b) 55

(c) The age is in the range 70–80.

## Section F (p 77)

**1** (a)

| Weight in kg | Mid-interval value | Frequency | Mid-interval value × frequency |
|---|---|---|---|
| 50–60 | 55 | 8 | 440 |
| 60–70 | 65 | 16 | 1040 |
| 70–80 | 75 | 10 | 750 |
| 80–90 | 85 | 6 | 510 |
| Totals | | 40 | 2740 |

(b) 68.5 kg

**2** $4210 \div 77 = 54.7$ minutes to 1 d.p.

**3** $925 \div 100 = 9.25 \,\text{cm}$

**4** (a) 20–30          (b) 85

(c)
| Age in years | Frequency |
|:---:|:---:|
| 0–10 | 10 |
| 10–20 | 25 |
| 20–30 | 30 |
| 30–40 | 20 |
| Total | 85 |

Mean = $1875 \div 85$
= 22.06 years to 2 d.p.

## Section G (p 78)

**1** (a)  9 years     (b)  10.5 years

(c)  $315 \div 24 = 13.125$ years

(d)  The mode, or median, because most of the people on the coach are about this age.

**2** (a)  Mean £1.17 to nearest penny
Median 20p
Mode 10p

(b)  The pupil's choice with reason

**3** (a)  Ranges     British: 20 cm;
                    Japanese: 19 cm

(b)  Medians  British: 147 cm;
                    Japanese: 141.5 cm
        Means     British: 147.3 cm;
                    Japanese: 142.25 cm

(c)  The Japanese students are smaller than the British students although the British students have a greater range in heights.

# Review 3 (p 189)

**1**  B costs £13.50,    C costs £5.40.

**2** (a)

Spinner B

|         |   | 0 | 1 | 2 | 3 | 4 |
|---------|---|---|---|---|---|---|
|         | 0 | 0 | 1 | 2 | 3 | 4 |
| Spinner A | 1 | 1 | 2 | 3 | 4 | 5 |
|         | 2 | 2 | 3 | 4 | 5 | 6 |
|         | 3 | 3 | 4 | 5 | 6 | 7 |

(b)  $\frac{3}{10}$

(c)

Spinner B

|         |   | 0 | 1 | 2 | 3 | 4 |
|---------|---|---|---|---|---|---|
|         | 0 | 0 | 0 | 0 | 0 | 0 |
| Spinner A | 1 | 0 | 1 | 2 | 3 | 4 |
|         | 2 | 0 | 2 | 4 | 6 | 8 |
|         | 3 | 0 | 3 | 6 | 9 | 12 |

(d)  $\frac{11}{20}$          (e)  $\frac{13}{20}$

**3** (a)  $4x - 11$          (b)  $3x + 3$
  (c)  $8x - 1$          (d)  $5x - 13$
  (e)  $9x + 5$          (f)  $11x - 5$

**4**

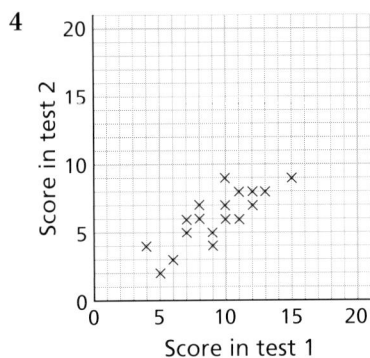

There is a fairly strong correlation between the two scores.

**5** (a)  $x = 24°$, $y = 18°$
  (b)  $z = 45°$

**6** (a)  130 m.p.h.
  (b)  97.5 miles
  (c)  5:42 p.m.

**7**  $v = 0.155$

**8**

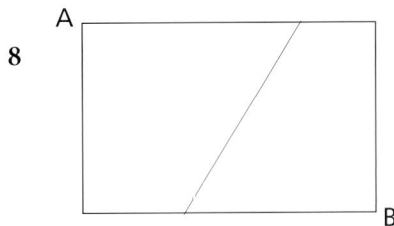

**9** (a)  23          (b)  44          (c)  49.5
  (d)  The English marks have a 'peak' in the 50s, but the maths marks have a 'peak' in the 30s.

**10**  This is one way to form the equation:
  Let $n$ be the number of rabbits Paul has.
  So Peter has $2n$ and Mary has $n - 31$.
  So $n + 2n + n - 31 = 161$
  $\qquad 4n - 31 = 161$
  leading to $n = 48$
  Paul has 48 rabbits, Peter has 96 and Mary has 17.

**11**  45

**12**  39%

**13**

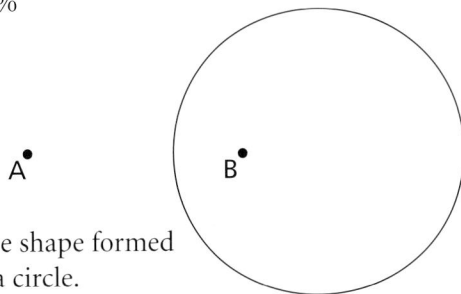

The shape formed is a circle.

**14**  $\frac{33}{48}$  or  $\frac{11}{16}$

**15** (a)  819          (b)  40–49
  (c)  4.5          (d)  44.7 (to 1 d.p.)

**16** (a) $-\frac{1}{3}$ (litres per kilometre)
  (b) $y = -\frac{1}{3}x + 80$    (c)  10

**\*17** (a)  $129 - n$    (b)  $5n$    (c)  $2(129 - n)$
  (d)  $5n = 2(129 - n) + 85$
  (e)  $n = 49$, so she has 49 5p coins and 80 2p coins

**\*18** (a)  300          (b)  4000          (c)  $n \times 10^{n-1}$

**Mixed questions 3** (Practice booklet p...)

**1** (a) 7:9

(b) 9:7 The ratio has been reversed

**7** 76.5 m.p.h.

**2** (a) 180

(b)

| Ordinary | 30-sided |
|----------|----------|
| 2 | 1 |
| 3 | 1 |
| 3 | 2 |
| 4 | 1 |
| 4 | 2 |
| 4 | 3 |
| 5 | 1 |
| 5 | 2 |
| 5 | 3 |
| 5 | 4 |
| 6 | 1 |
| 6 | 2 |
| 6 | 3 |
| 6 | 4 |
| 6 | 5 |

(c) $\frac{1}{12}$

**3** $x = 80$

**4** (a) 17    (b) $^-1$

**5** The pupil's locus and point

**8** (a)

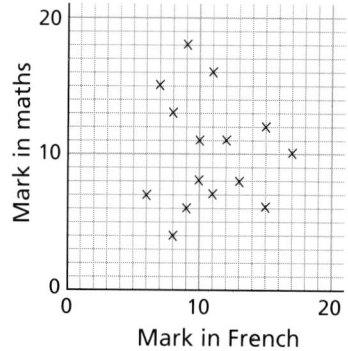

(b) There is no correlation between the two scores.

**9** £27.50

**10** $\frac{13}{25}$

**11** CD album 24 euros, carry case £8.00

**12** 97 cm (to 2 s.f.)